Chasing Dragonflies

ALSO BY CINDY CROSBY

Tallgrass Conversations: In Search of the Prairie Spirit
(with Thomas Dean)

The Tallgrass Prairie: An Introduction

Chasing Dragonflies

A Natural, Cultural, and Personal History

Cindy Crosby

Illustrated by Peggy Macnamara

NORTHWESTERN UNIVERSITY PRESS
EVANSTON, ILLINOIS

Northwestern University Press
www.nupress.northwestern.edu

Kobayashi Issa, "Days Are Short," on page xvii, from *Dragonfly Haiku*. English version copyright © 2016 by Scott King. Reprinted with permission from Red Dragonfly Press.

Dennis Paulson, "The Miraculous Rectum," on page 25, from *Argia* 30, no. 1 (March 15, 2018). Reprinted with permission from Dennis Paulson.

W. S. Merwin, "After the Dragonflies," on page 60, from *Garden Time*. Copyright © 2016 by W. S. Merwin. Reprinted with the permission of The Permissions Company, LLC on behalf of Copper Canyon Press, coppercanyonpress.org.

Scott King, "The First Dragonfly of the Year," on page 177, from *Dragonfly Haiku*. Copyright © 2016 by Scott King. Reprinted with permission from Red Dragonfly Press.

Ken Tennessen, "Spring Comes," on page 181, from *Dragonfly Haiku*. Copyright © 2016 by Ken Tennessen. Reprinted with permission from Red Dragonfly Press.

Printed in the United States of America

10   9   8   7   6   5   4   3   2

Library of Congress Cataloging-in-Publication Data
Names: Crosby, Cindy, 1961– author. | Macnamara, Peggy, illustrator.
Title: Chasing dragonflies : a natural, cultural, and personal history / Cindy Crosby ; illustrated by Peggy Macnamara.
Description: Evanston, Illinois : Northwestern University Press, 2020.
Identifiers: LCCN 2020001370 | ISBN 9780810142305 (paperback) | ISBN 9780810142312 (ebook)
Subjects: LCSH: Dragonflies. | Dragonflies—Ecology. | Gardening to attract wildlife.
Classification: LCC QL520 .C76 2020 | DDC 595.7/33–dc23
LC record available at https://lccn.loc.gov/2020001370

For Jeff, who chases dragonflies with me

It seems to me that the natural world is
  the greatest source of excitement;
  the greatest source of visual beauty;
  the greatest source of intellectual interest.
It is the greatest source of so much in life
that makes life worth living.

—DAVID ATTENBOROUGH

CONTENTS

Dragonfly and damselfly common names have been capitalized to be easily recognizable as such. Scientific names are used only on first reference in the book and are italicized. Family names for Odonates are capitalized, but the common name, instead of the scientific name, is usually given. Other common names of living things (plants, mammals, birds), such as "red-winged blackbird" or "pasque flower," are not capitalized, nor are their scientific names given. Endnotes, instead of footnotes, are used throughout. These stylistic choices have been made for ease of reading and understanding on the part of the reader.

Prologue

It might have been otherwise.

—JANE KENYON

In my mailbox is a birthday card from my brother Chris. I open the envelope and shake out a wooden bookmark etched with the image of a dragonfly's wing. I'm touched. But not entirely surprised. To see anything dragonfly-related prompts my family to think of me. This connection to dragonflies has come about gradually over the past fifteen years—so gradually that some days I'm not quite sure how it all happened. But there it is. Dragonflies and my life are intertwined.

That day, I had returned from Florida, where I celebrated my birthday with my husband, adult children, and six assorted grandkids under age eight. It was the perfect week. Beach, pool, family cookouts, more beach time. Southern dragonflies buzzed through the landscape, many new to me. Interesting shorebirds. Exotic flowers in tropical colors. I felt happy. At peace. Rested.

Then, as our vacation was about to end, my phone vibrated. I saw there was a message from my doctor's office. Opened it.

The word stared up at me.

Malignant.

Before I'd left home, I'd had a few tests. The doctor who did the first test, and then the second, was reassuring. *I've done thousands of these biopsies! This one doesn't look problematic. But of course we won't know until the tests come back.*

And now they had.

"There is no reason to write a book unless the process of imagining it changes one's life forever," asserted Richard Manning in his opening chapter of *Grassland.* This diagnosis dropped into a life molded and shaped by the rhythms of dragonflies. From April to the end of October, I'm restless whenever I'm indoors, wondering what I'm missing by not being outside. The life of the skies and the watery underworld of creeks and ponds have me in their grip; I don't want to ever feel it loosen.

Dragonflies have become a part of my identity. They've been a solace through my cancer diagnosis and my recovery as I watched them fly through my backyard, where I was relegated to a lawn chair in the midst of writing this book and unable to chase them. Dragonflies were also an introduction to hundreds of kind and interesting people, many willing to share their passion for dragonflies and Odonates by phone and email and by my side in the field, mentoring a willing learner like myself. Dragonflies are my nemesis too, the cause of frittering away hundreds of hours of time that might have better been spent earning a paycheck. Or so some people might think.

Once you "see" dragonflies, your world will change. Every backyard barbeque, each walk along a river, or time spent weeding a garden—suddenly, you notice—they are everywhere! So many aloft. So many fluttering in the grasses, skimming ponds, hovering around traffic lights. Each one a bit of unique insect art.

In *An Obsession with Butterflies*, author Sharman Apt Russell says that adding butterflies to your life is like adding another dimension: "All this existed before, has always existed, but you were unaware. You didn't see." Russell's butterflies have been my dragonflies.

Dragonflies say "mystery" to me.

There is so much we don't know about the order Odonata. So much to learn. Dragonflies live most of their lives under the surface of the water. One fine day, they pull themselves out, split their old "skins," and, changeling-like, become something beautiful, colorful, and new. Grow wings. Take to the air. Their brief lives are over before you can blink. Or so it seems. As Kobayashi Issa, the Japanese haiku poet, wrote:

Days are short—
the dragonfly's life
fleeting, as well

The first half of my life, I had a lot of pat answers to some of life's most difficult questions. The second half of my life I live knowing some answers are going to be in short supply. Cancer, with its shattering shock waves and mysteries, reinforced this realization. The dragonflies, with their ancient lineage and predictable lives—yet shot through with mystery and the unknown—echo this enigma. Their lives are tenuous, as our lives are. Dragonflies move between water and air, transforming themselves, all while prone to the whims of weather and the vagaries of the next frog, bird, or other predator waiting to snatch them from life.

And yet, for hundreds of millions of years, they have been evolutionary survivors.

As I write these words, a pandemic is sweeping the globe. Illinois families are sheltering in place to avoid becoming infected with the COVID-19 virus. We're unable to go about the normal rhythms of our days. Workplaces, nature centers, houses of worship, schools—all are closed. My husband, Jeff, and I drive to our daughter's house and talk to the kids from our car, while they stand on the porch. I teach my classes online. Fear, anxiety, and uncertainty reign.

And yet. Temperatures warm. The month of April arrives. My crocus and daffodils bloom. I stand on the back porch and scan the skies. Although my life and millions of other lives are in complete disarray, the dragonflies are unaffected. The first migrants will arrive any day. Their rhythms of life go on. I find comfort in this. We don't know how this global pandemic will end. But there is solace in the rhythms of the natural world.

I'm inspired by how dragonflies are both tough and fragile; fierce and mild. They each fly for only a few short weeks, yet the species is still around after others have disappeared from the earth. They are cannibals who may eat one another, yet you can safely hold one in your hand. As they transform themselves from water creatures to creatures of the air, they are vulnerable to the predation of frogs or birds. A falling leaf may damage their newly unfolded wings beyond repair. Yet when those same wings harden, they are strong enough to carry many of them thousands of miles in migration. These dichotomies seem to hang in opposition, yet they are all part of what makes a dragonfly the fascinating creature that it is.

There are so many reasons to fall in love with dragonflies. They have intriguing stories to tell us. They make our world a more beautiful place. And they are scrappy survivors.

Writing this book has been a reminder of how dragonflies have become inseparable from who I am. They've taught me to slow down. They were a healing presence in my backyard as I recovered from cancer surgery. They are a welcome distraction and a reminder of the normalcy of life in the natural world during the uncertainty of a pandemic. They've helped me learn to pay attention more closely to the natural world and to my life. Dragonflies delight and astonish me. Sharing them with you is part of my journey.

Let's go chase some.

Chasing Dragonflies

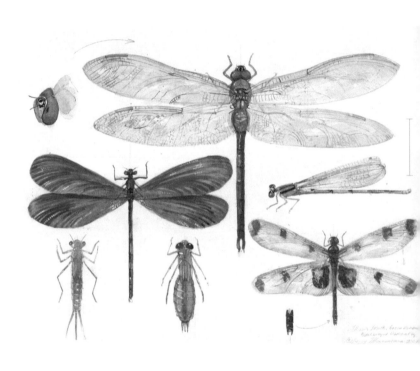

Why Chase Dragonflies?

All about Odonates

Today I saw the dragon-fly
Come from the wells where he did lie.
An inner impulse rent the veil
Of his old husk; from head to tail
Came out clear plates of sapphire mail.
He dried his wings: like gauze they grew;
Thro' crofts and pastures wet with dew
A living flash of light he flew.

—ALFRED LORD TENNYSON

Why should anyone care about dragonflies—and damselflies, their kissing cousins? Do we really want more insects in our yards? What motivates people like you and me to tramp through wetlands, waving off mosquitoes and blackflies; to make hash marks on a clipboard, noting the names of the species we see and the numbers we observe? Why, in a world full of pandemics, war, poverty, political unrest, and other serious issues, does chasing a "bug" matter?

Many people know what a dragonfly looks like, but few of us really notice them as we go about our business of living. Dragonflies and damselflies are short-lived creatures who spend more of their lives under the water than they do in the air. Most of us who do mark their presence from time to time know them as beautiful denizens of the sky, not the ferocious predators of pond and stream.

Despite their brief time in the world, dragonflies have a long and rich cultural history. They play important roles in religion, art, music, poetry, and literature. Their unusual sex lives are unparalleled by any other living being in our natural world. A dragonfly's incomplete metamorphosis from creature of the dark to graceful navigator of light and air holds symbolism for people from diverse cultures and backgrounds.

Monitoring these creatures—chasing dragonflies—over time may give us important information about how our planet's climate is changing. When we regularly go out and count their numbers and species in a given location from year to year, we learn what happens to insect and aquatic populations when we develop wetlands, pollute a river, or change the character of a river or stream. What we discover about dragonflies informs the future management of prairies, marshes, bogs, woodlands, and wetlands. Dragonflies are a barometer of the health of the world we are leaving to our children and grandchildren.

Because of their amazing assembly of body parts, dragonflies also hold clues for human air travel, and even drones. And did I mention the number of mosquitoes dragonflies eat?

If you are conservation-minded, you already know that insect populations are in steep decline. The phrase "insect apocalypse" has even been bandied about. *Fewer pesky bugs*, you might say? Isn't that a good thing? Think about it. The insect world is a building block upon which life depends; a food source which many creatures need for survival. It includes many of the pollinators who, through their work, give us delicious food to eat and refreshing beverages to drink. Sure, dragonflies are not pollinators. But they are an important part of the biodiversity that makes up a healthy world.

Dragonflies changed my life. But I wasn't always "present" to them. I hadn't really noticed dragonflies or paid much attention to "Odes" until I moved with my husband, Jeff, to the Chicago region when I was in my late thirties. I was not enamored of the frantic busyness of the Chicago suburbs where we lived and struggled to put down roots. The hundred-acre prairie down the road from my house was a place I often walked, journaled, and tried to make sense of where my life was headed.

An old wooden bridge over a brook there became my favorite front-row seat for the life of the tallgrass prairie, a landscape which intrigued me, and which—at that time—I knew almost nothing about. Afternoons and evenings, I sat with my journal. Wrote a little. Sketched a bit. But mostly observed the life of the prairie unfolding around me.

On a summer morning, as I relaxed against the warm boards of the bridge and watched the water move downstream, I noticed clouds of black-winged insects looping in and out of the shoreline vegetation. At that time, I was in a period of naming things. I wanted to know who I was looking at, and something about their natural history. I knew they were insects. *Butterflies?* Perhaps. I sketched the "black butterflies" and then scrambled down the banks to get a closer look. I thumbed through the field guides I'd brought along to try to figure out what I was seeing.

Finally, it hit me. These were no butterflies. These were damselflies, a type of insect I knew almost nothing about. Specifically, these were Ebony Jewelwing (*Calopteryx maculata*) damselflies, with iridescent metallic green bodies and raven wings. I learned they are closely related to dragonflies, which was something I could put a name to.

I'm sure I grew up seeing dragonflies and damselflies. My mother made sure we spent a lot of time around the water. My maternal grandmother was an elementary school science teacher, and she helped me name trees and birds I didn't know. My grandfather built a boat and took my brother, my sister, and me fishing every summer. I spent many hours in and around lakes, streams, and ponds.

Doubtless, there were dragonflies everywhere. I remember seeing frogs as a child. Birds? Of course. I chased butterflies around an empty lot just down the road from our house and marveled at the variety of ones I would find. But I've lost any memory of dragonflies.

My brain had filtered them out . . . until now. Little did I know what this introduction would mean. As a citizen scientist who now monitors dragonfly and damselfly populations, the "dragonfly calendar" is a part of my yearly planning. My life would become attuned to the rhythm of their emergence, flight, and disappearance each year. I'd chase them on my travels to other countries. I'd comb bookstores for new books; search online for new species. You might say dragonflies became a minor obsession.

Okay, maybe more than minor.

But that long-ago day, as I explored the field guides looking for the "black butterflies," I woke up to the world of Odonates. *Dragonflies? Damselflies?* Different, but similar.

7

What's a Dragonfly?

It's not unusual to confuse damselflies with butterflies at first, as I did. An even more common confusion is mixing damselflies with dragonflies. Often, many of us talk about *dragonflies* as an inclusive term that embraces their more ethereal kindred, the *damselflies*. I do it a lot. But are they the same? No.

So what's the difference?

Dragonflies and damselflies are both part of the order Odonata, which simply means "toothy ones" or "toothed ones." They have these nicknames because of their jaws. Each has four jaw parts: the *labrum, mandibles, maxillae,* and *labium.* The mandibles have *sclerotized,* or hardened, "teeth" referred to as their *molars* and *incisors.* This mouth arrangement allows a dragon or damsel to catch prey on the fly. If you see a dragonfly up close, you'll understand one of the reasons dragonflies and damselflies have reputations as fierce and tireless predators. They are carnivorous. The insect's mandibles are often stuffed full of a luckless moth or small bug. Or even another dragonfly or damselfly.

While not the same, dragonflies and damselflies are closely related. Simply put, they are part of an insect order. Remember high school biology, where you learned "kingdom, phylum, class, . . ." and so on, down through "family, genus, species"? Channel that time in your life for a moment. Insects are in the class Insecta, or class Hexapoda, if you prefer. I like to refer to both dragonflies and damselflies collectively as Odonates, or Odes, as they are in the order Odonata. The insect order Odonata may be broken down in North America into two simple "subgroups"—the Anisoptera (dragonflies) and the Zygoptera (damselflies). There are suborders and infraorders, but for our purposes, these are the two to keep in mind.

This classification is a relatively modern concept, as Edwin Way Teale pointed out in his 1960s classic *The Strange Lives of Familiar Insects*. At one time, earthworms and even snakes were lumped in with insects by naturalists, Teale explained. As technology tells us more about how each insect, plant, or mammal fits into the natural world and their relation to one another, the finer classifications continue to be shuffled around.

Dragonflies are not the oldest insects in the world (at this writing, that honor goes to the Devonian *Rhyniognatha hirsti*, at approximately 400 million years old from fossil records), but we do know that the ancient relatives of the dragonflies, the Protodonata, or "griffinflies," were flying before the age of the dinosaurs, perhaps 250 to 325 million years ago. Dragonflies are one of the oldest flying insect groups in the world. They are virtually unchanged today, except in size. At one time, their ancient relatives were almost two-and-a-half feet long, from the face to the tip of the abdomen, with wingspans of up to two feet or more. Scientists speculate that the more oxygen-enriched atmosphere allowed insects to be larger.

Think about it. Imagine a swarm of those giant insects flying toward you. Alfred Hitchcock's movie *The Birds* would have nothing on this mob, intent on their prey. Today, most of us in the Midwest will only experience a swarm of Common Green Darners (*Anax junius*) flying around us, with three-inch-long bodies. Quite a difference.

These dragonfly "chassis" are vulnerable, yet strong. A dragonfly might be snapped up in an instant, its life snuffed out by a bird or frog. But their bodies are tougher than they look, and they have served dragonflies well, with little variation, for millions of years.

Let's take a closer look at dragonfly anatomy and why it matters.

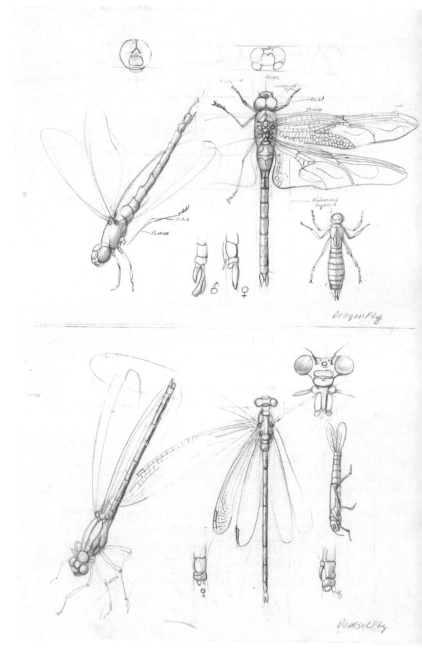

Dragonfly

Damselfly

2

What's a Dragonfly, Anyway?

An Anatomy Lesson

The starting point must be to marvel at all things,
even the most commonplace.

—CARL LINNAEUS

When I give a talk on dragonflies, people often show me cell phone photos of "dragonflies" they ask me to identify. Many of the insects are look-alikes and not dragonflies at all. They are mayflies. Antlions. Owlflies. Or some other insect. I used to make the same mistakes myself. How can you know what is and isn't a dragonfly?

Let's start with what dragonflies are *not*. Dragonflies and damselflies are often called bugs. In fact, they are not "true bugs," which are an order of insects known as Hemiptera. True bugs have sucking mouthparts, and this order includes such insects as cicadas and leafhoppers. Although not "true bugs," both dragonflies and damselflies are insects, so they have the requisite insect parts: a head, thorax, and abdomen; six legs, four wings, and two antennae.

Perhaps as a child you sang "The Insect Song" with its enthusiastic hand motions. I like to sing a variation of it in my adult dragonfly ID classes as well, specifically adapting it for dragonflies and damselflies and their ten abdominal segments:

> *Head, thorax, abdomen (one to ten)*
> *Head, thorax, abdomen (one to ten)*
> *Six legs, four wings, two antennae*
> *Head, thorax, abdomen (one to ten)*

When I give workshops on dragonflies, I've found that members of the Illinois garden clubs—and quite a few men and women of Audubon and Wild Ones native plants groups—are great singers. They are also game to do the hand motions as they belt out the lyrics with me. Once you sing the song, you'll know the basic body parts of a dragonfly and damselfly.

Why learn the parts of a dragonfly? Knowing how an Ode is put together is crucial if you use a field guide to try to ID dragonflies or damselflies. Anatomy matters.

When I first received my cancer diagnosis, my surgeon spent two hours explaining how different parts of my anatomy functioned, which parts were affected, and what would be surgically removed. I was shocked at how little I really understood about the body that had served me so well for more than fifty years. Who knew lymph nodes were so important? I was also dismayed at how my body had "betrayed" me. Without this understanding—knowing my anatomy and how each body part functioned—I couldn't fully comprehend my surgery and the ramifications for my life and future going forward.

In an age where we can Google anything we want to learn, it's important to remember there are some facts we need to memorize for ourselves. Ode anatomy is one of those building blocks dragonfly chasers need to know. If you truly want to "see" dragonflies and understand how they function in your backyard and your world, it helps to learn their basic anatomy and how each body part works. Then, you'll have more confidence in identifying a dragonfly or a damselfly and understand more about its habits and habitats. Let's dive deeper.

To recap the song lyrics: All dragons and damsels have a head, a thorax—what we'd think of as our chest—and an abdomen, which is sometimes called a tail by those not in the know. "Six legs" distinguishes dragonflies and other insects from spiders and millipedes, which have more legs and aren't true insects. The six jointed legs are arranged in three pairs. Dragonflies and damselflies both have two pairs of wings, but while the damselfly's two wing pairs are pretty much the same shape and size, the dragonfly's hind wings are larger than their forewings. *Anisoptera* (the name of the dragonfly infraorder) means "unequal wings" or "different wings." *Zygoptera* (the name of the damselfly suborder) means "even wings" or "same wings."

Got that? Let's move to the abdomen, that skinny long part of the dragonfly and damselfly, which is divided into ten segments. Each of those segments contains markings—dots, dashes, arrows, and long stripes—that help us determine the kind of dragonfly (the species) we're looking at. The abdomen also contains the reproductive organs, so you can figure out male or female if you're interested. Some Odes are sexually *dimorphic*—that is, the males and females of the same species may *look* different from each other in coloration and pattern.

If you've ever tried to identify a dragonfly or damselfly on the wing, you know how difficult it can be. Many dragonflies and damselflies look superficially similar. The difference between one species and another might be a yellow spot on abdominal segment three or a particular blue stripe pattern on the thorax. Knowing Ode anatomy will help you put a name to a dragonfly when you use your field guide or phone app.

Back to those legs. Dragons and damsels have three pairs of jointed, bristly legs, specially equipped for holding insects as they eat. A bug basket. There is a last segment with a "foot" on it, which is actually part of the leg and has two hooks. This is called the *claw*. But despite all these great leg parts, dragonflies don't really walk. Although they are proficient fliers and can move up and down plant stems, you won't see a dragonfly adeptly strolling across a horizontal surface.

Let's return to the head. The antennae are present, but they are pretty much invisible unless you're up close. Ode experts believe the antennae are important in detecting air movement and may help regulate the dragonfly's body orientation during flight.

But don't look for a nose on a dragonfly or damselfly. It doesn't exist. Older reference books say dragonflies have no sense of smell, but new research shows that dragonflies may use sensory pits on the antennae, which are believed to play a role in "smelling" and detecting prey. What about hearing? Scientists at this writing don't believe that dragonflies and damselflies can hear us, in the way we think of hearing, or vocalize like a bird or a frog.

So dragonflies and damselflies—Odes—are alike in many ways. But then, the differences begin.

Distinguishing between
Dragonflies and Damselflies

How do you know whether the insect you are looking at is a dragonfly or a damselfly?

When I hike the prairie trails or a path by a stream or pond, damselflies tend to be fluttering around me at about knee height. I usually don't see them high in the air, as I do dragonflies. That's a good first clue.

Dragonflies and damselflies also look a bit different in the air. Damselflies tend to have a weaker flight pattern and look slimmer and more ethereal. In fact, the word *damselfly* is derived from the French word *demoiselle*, meaning "young mistress."

Dragonflies are often—but not always—chunkier, robust, and more muscular fliers. Remember, we're talking in generalities here. There are going to be exceptions. The Calico Pennant (*Celithemis elisa*), which is one of the most beautiful midwestern dragonflies in my estimation, and the Halloween Pennant (*Celithemis eponina*) are both tinier than you'd expect. You're more likely to see them perched on the tips of plants than flying overhead like their stouter cousins in the Emerald or Darner families.

So how are you going to know it is a dragonfly? By its wing position—flat and open at rest. Like an airplane. When perched, most damselflies fold their wings together. And, as we said earlier, dragonflies also have larger hind wings than forewings, unlike damselflies, whose wing pairs are roughly similar.

A damselfly's eyes are set on each side of its head, often compared to a hammerhead shark's eyes. In contrast, dragonflies—with some exceptions—have eyes that take up much of their heads and touch, or nearly touch. So, if you can see the eyes, you can usually make the distinction. Eyes close together? Dragonfly. Eyes apart? Damselfly.

Let's talk about those eyes for a moment. The eyes of a dragonfly have up to thirty thousand visual units honey-combed together—one of the many reasons it is difficult to sneak up on a dragonfly. Damselflies have five thousand to ten thousand individual light-sensitive units. A dragonfly's eyesight is keen, and a dragonfly's color vision is said by some experts to be the best in the world. Dragonflies can also see ultraviolet rays and detect polarized light.

Three other "simple eyes" are triangulated on the top of the head in both dragonflies and damselflies. These simple eyes, the *ocelli*, function to give the dragonfly visual input that helps it maintain flight stability and balance.

Exceptions, Exceptions, Exceptions

Of course, there are exceptions. Science wouldn't be nearly as much fun without them. Eyes? Some dragonflies—for instance, the Clubtails—have eyes that are set apart. Wings? Spreadwings in the family Lestidae are the rebellious teenagers of the damselfly world. Instead of folding their wings together, like every damselfly species is supposed to, they hold their wings at a forty-five-degree angle when at rest. I imagine them singing along with Sammy Davis Jr.: "Me— I've gotta be me!" Ode expert Dennis Paulson tells me this wing position may help them quickly jump into flight to snag a passing insect.

There are more exceptions to these general distinctions, but if you know these important clues—resting wing position, general body form, flight type, and eye position—you can figure out the others as you go with a good field guide.

Learning dragonfly anatomy is a way to learn to pay close attention to the natural world. As you study an Ode at rest in the field, at home from a photograph, in the hand, or aloft on a summer's day, you'll develop appreciation for its complex yet simple machinery. We've touched on the basic body parts, but there are many more delightful anatomical features to learn: the stigmas on the wings, names for the parts of the face, the reproductive components of each Ode. Each part of a dragonfly's or damselfly's anatomy holds clues to unlocking its identity, and its habits.

As I learned more about Ode anatomy, I felt a surge of gratitude for the intricacy of these flying machines. I found that the close examination of dragonfly anatomy gave me a fresh appreciation for other insects, and a desire to know more about them. One of the benefits of nature study!

Learning dragonfly anatomy may also have unanticipated rewards. One of my dragonfly workshop students recently emailed me to say he won a pub trivia competition. His win hinged on a question about the difference between dragonflies and damselflies.

Who knows? Maybe you'll astonish your friends and amaze your family at the next trivia gathering by pulling out these factoids. But besides a possible *Jeopardy!* Daily Double win, there are many other good reasons to learn about dragonflies and then go chase them. Their unusual life cycle is another such reason. Let's investigate.

3

Embracing Change

Transformation

Change is inevitable—except from a vending machine.

—ROBERT C. GALLAGHER

Such a peaceful morning, paddling through the cattails. Great blue herons and egrets flap over my head. Marsh wrens sing. I'm kayaking Wisconsin's Horicon Marsh with my husband, Jeff, and our friend Mary. Dragonflies and damselflies zip across the marsh, defying the stiff wind.

Then, I see her. Alongside the kayak, an unknown damselfly is struggling. Perhaps she's been blown off the vegetation that she used to pull herself out of the water. She's still in the *teneral* stage—somewhere between life as a newly emerged damselfly nymph and an adult; ready to shed her old life of the underwater world and trade it in to become a brightly colored creature of the air. But this one is doomed to drown before she ever takes flight.

I'm not usually a meddler, but I fish the damselfly out of the water with my paddle. What is the life of one damselfly, more or less, in a marsh that is teeming with them? *And yet.* Everywhere this morning, I've seen flashes of blue and tan damsels, busy with the charge of repopulating the marsh. I place the half-drowned insect on my knee in the sunshine to dry off and continue paddling to catch up with the others.

The brisk wind rumples the damselfly's flimsy wings. I lower my leg deeper into the kayak to give her shelter from the breeze and look more closely. She's pale beige. Species? Impossible to tell. Too early. Her abdomen slowly uncurls, then her wings straighten. She uses her forelegs to wipe her face and then, looks up. I swear I can feel the damselfly's beady gaze taking my measure.

She rides shotgun, mostly motionless, as I paddle through the marsh. After an hour or so, the damsel suddenly flutters her wings and then—like a tiny, light-filled comet—lifts off and streaks into the cattails.

I watch her until she disappears. The kayak feels emptier. But the damselflies all around me seamlessly assimilate the new one; the cattail marsh absorbs one more insect.

We think of dragonflies as beautiful aerial acrobats in a kaleidoscope of bright colors. Most people are shocked to learn that dragonflies and damselflies spend most of their lives under the water as creepy—yet marvelous—critters.

The life of a dragonfly or damselfly is all about change. There are degrees of change—from egg to nymph; from nymph molting to bigger nymph, repeatedly molting to bigger nymph; and finally from nymph to the transcendent creature of the air, that motif we know from shower curtains, necklaces, T-shirts, and lampshades.

Let's begin with dragonfly eggs. After mating (more about this astounding topic to come), dragonflies *oviposit*, or lay eggs. I love walking my routes and seeing the diversity of ways different species accomplish this, from tapping eggs into the water, to slitting open streamside plants for an egg deposit, to studiously pushing eggs into floating vegetation mats. The eggs may hatch in as short a period as five days in temporary pools to overwintering in some species.

Dragonflies go through "incomplete metamorphosis" consisting of three stages: egg, nymph, adult. Most of us are familiar with butterflies' more highly evolved "complete metamorphosis" that involves egg laying, caterpillar hatch (larva), pupa or chrysalis, and then adult butterfly. Dragonflies skip the chrysalis stage and go straight from nymph to adult.

So out of the dragonfly eggs come not those glamorous flying creatures we admire aloft but rather some unusual-looking offspring. They've been called *naiads*, *nymphs*, and *larvae*, with *nymphs* being the most widely used term in North America. Think of a ferocious-looking beetle in greens, tans, blacks, or browns. These are the creatures you'll see schoolchildren scooping from the water with their aquatic nets into trays, learning the ways of underwater life. Or dragonfly chasers, netting nymphs, trying to learn to tell apart the different dragonfly and damselfly species in the early stages. Not easy.

A nymph moves through the water by expanding and contracting its rectum internally, letting water pulsate over the gills in its branchial basket–in other words, blowing water out its butt. You heard that right! This method of propulsion gives a nymph a quick, rocket-like boost toward prey. It's a miraculous natural history description that will endlessly captivate any class of third graders. Adults are riveted by the description as well. Ode expert Dennis Paulson captured this natural history phenomenon in his poem "The Miraculous Rectum":

> Nobody loves the rectum; at best they say "ew."
> If you try to correct 'em, you convince very few.
> But this part of the body, so important to the nymph,
> is something to admire, so don't just say "hmph."
> The rectum in a nymph, surprising as it seems,
> does more than just digestion; it has other schemes.
> This miraculous organ has three different uses,
> and they function alike in all ode cabooses.
> Where digested remains have just gone their way,
> fresh water comes into it, saving the day.
> Oxygen's taken up by tracheae tiny,
> and CO_2 purged through the opening spiny.
> So respiration's served, but "What else?," you ask.
> There's also locomotion for still another task.
> The abdomen's compressed, the water goes south,
> and the larva shoots out from the predator's mouth.
> So if anyone asks you where to direct 'em
> for dragonfly miracles–point to the rectum.

But are these nymphs comic figures? No. These are fearsome predators of the deep. Their jaws are comparable to a spring-loaded hinge, known as a "distensible" hinge. As they eat whatever crosses their underwater path—mosquito, midge, and mayfly larvae; tadpoles, small fish, other dragonfly nymphs—they shed their exoskeleton, or molt. This process can take place from eleven to fifteen times during the underwater portion of a dragonfly's life. This may seem like a "whole lotta moltin' goin' on," but consider in comparison the mayfly, which is said to molt more than fifty times. And *we* think change is hard!

The outer shell that the dragonfly or damselfly nymph outgrows is made partially of a substance called *chitin* and is known as the *cuticle*. Its exoskeleton is made up of layers of cuticle and waxes. When the new exoskeleton is ready, the outer, outgrown one splits open to allow the nymph to continue growing. The new cuticle is soft and flexible and expands upon breaking through the old skin; within a short time, however, it too hardens and no longer can expand. And so it goes: on to the next molt.

As Odonate nymphs go through successive molts, you'll see an increase in the lenses of the eyes, new joints in the antennae, and the addition of wing pads on the thorax. Nymphs go about their business in their underwater world for as little as several months to as long as seven years. Altitude and latitude, as well as water temperature and the length of a growing season, are believed to play a role in determining the length of time spent in this stage.

Observes Edwin Way Teale, "Such changes are the prelude to the greatest and most dramatic change of all, the transformation from the muddy, water-world nymph to the colorful, glinting, winged creature of the air." In the teneral stage, the nymph sheds its "skin," or exoskeleton, and pulls its adult dragonfly body with wings from its old "shell." *Hemolymph*—insect blood—pumps through the wings, enlarging them into the strong flight instruments they will become.

The adult dragonfly has now stopped growing. It is the size it will be for the rest of its life. As an adult, that Ode's life will be brief: maybe a few minutes if a hungry bird is nearby. Or a few weeks—even months—if it is lucky. In some species, you'll notice age by the degree of *pruniosity*—bluish powdering—on the body. Other older dragonflies and damselflies often have tattered wings, torn by predators.

Transformation

One morning I kayak the lake at Busse Woods, a Cook County Forest Preserve close to my home, and watch the teneral dragonflies emerge along the shoreline. They are pale; so very pale. It's impossible for me to see what species they will become. What potential! I marvel over this scene, reenacted again each day. And yet, it mostly goes unnoticed. So much change.

The writer Robin Wall Kimmerer tells of a note she saw on a coffee shop tip jar: "If you fear change, leave it here." Much of my life has seemed to be about becoming something new. There is tension between the person I am and the person I want to be. In my twenties, my life focused on my children, the garden, and the independent bookstore my husband and I ran. In my thirties, it was about being a news journalist and book reviewer, where life revolved around deadlines and interviews and being "busy." In my forties, I became a national park guide and then, briefly, a national park ranger, and I learned I could solo backpack thirty miles or more, drive a boat, and push the limits of my physical, mental, and emotional endurance.

Now, in my late fifties, I've found that age brings greater acceptance of change. You discover, once again, the value of close relationships. You accept what you can do and what you cannot. You learn to let go. You find you can live through "dark nights of the soul" and come out on the other side—like a long-lived dragonfly—bedraggled, tattered, but still *you*.

I watch one of the newly emerged dragonflies spread its wings to dry and harden in the sunshine. This is one of the most vulnerable times in a damselfly's or dragonfly's life. Any passing frog, bird, or fish can easily pick it off. Ode expert Kurt Mead says that as many as 90 percent of the tenerals may be eaten by birds at this stage; ants and spiders also take their fair share. And, as Teale reminds us, "At this point, if the helpless adult should fall into the water—which has been its home for many months—it would drown. It has ceased to be a water-breathing creature."

The nymph may emerge just before dawn to lessen its chance of getting eaten. When the wings are ready, the adult dragonfly lifts off into the sky, leaving its old "skin"—the *exuvia*—behind, often still attached to a blade of grass along a stream or pond.

What a symbol! This transformation—passing from a mostly unnoticed creature of the gloomy depths to the sparkling jeweled pilot of the skies—is one of the reasons people are so attracted to the dragonfly. Sometimes, we yearn to leave our old selves behind. It is this symbol of emergence and flight that has encouraged many of us through depression, loss, poor health, or times of creative or professional stagnation.

The idea of a new self—one with superpowers, so it seems—has always been enticing to me. It became even more seductive after my cancer diagnosis. What would it mean, to shed my old, ill, unwieldy body and become something fresh, healthy, and different?

Back to our dragonfly. Take a closer look at those brand-new wings in a photo or in the field. See the veins? These are the scaffolding. The thin wings are made of chitin, which also helps create the exoskeleton. It's a complex design. The way the wings are formed helps ensure stability during flight and helps the dragonfly perform those crazy aerial maneuvers. Each wing moves independently, allowing the dragonfly to fly up, down, sideways, and even briefly backward. Once they harden, the wings are tough. When I handle a dragonfly, it is always by holding it with the wings pressed together in order not to harm it.

These wings are inspirational in more than a spiritual sense. By observing and understanding how Odonate wings work, engineers learn more about flight. Think of drones, which hover, and move up, down, backward, and forward like a dragonfly. Drones are used in missing-person searches, so who knows what studying dragonflies might accomplish for people? A lovely example of biomimicry. Drones are also used by the military for spying—perhaps not such a lovely example—but hold promise for helping make the world a more peaceful place.

After learning about the dragonfly's life cycle, I found that I never look at a pond or a stream or a river—or even a puddle—in the same way. Dragonfly populations of various species are directly linked to how a stream flows, or to the substrate in the bottom of a pond, or to the disturbance of a watershed. Drill a well? Your dragonfly populations may change. Channel a stream? Ditto. When water quality and availability change, our dragonfly populations and species compositions may change as well. Sure—dragonflies as an order of insects are tough. They are long-term survivors. But development, pollution, and climate change may have an impact on them over time.

As I walk my routes at Nachusa Grasslands and the Morton Arboretum, I watch brand-new dragonflies picking off hundreds of mosquitoes. For most of us, knowing that dragonflies are out there performing a mosquito-removal service is sufficient reason to endear them to us. But even if they were here purely for our aesthetic pleasure, it would be enough. In a world full of hatred, political intrigue, cruelty, and natural disasters, we need dragonflies to add beauty. We need hope for something better.

The cycles of change in a dragonfly's life—without regard to political posturing, war, racial unrest, or the next mass shooting—are dependable all summer long. It doesn't matter if my personal life is in disarray. The dragonflies don't care about my health. They go about their business in complete disinterest. That knowledge gives me a certain peace.

Life seems like a match struck in the darkness—a brief flame—and then . . . gone. When my life ends, the dragonflies will continue their courtship dance. Spreadwings will slit leaves, drop eggs. Painted Skimmer (*Libellula semifasciata*) nymphs will squirt around the pond bottom, hunting their lunch. In creeks and streams, backlit by the early morning dawn, nymphs will continue their ancient breakthroughs of the water's surface. Then, compelled by some mechanism we only dimly understand, they will peel off their old bodies and pull out seemingly new creatures. From fragile to tensile strength. They will spend a few minutes or days or weeks aloft as glorious denizens of the sky. Then they will vanish, mostly unremembered and unremarked upon.

I think back to that damselfly at Horicon Marsh. In the context of this vast area—at thirty-three square miles, the largest freshwater cattail marsh in the United States—a damselfly smaller than the length of my pinky finger isn't much. At most, my teneral damselfly will live for several weeks. More likely, it will become fish food or a bird snack. If she does succeed in mating and laying eggs, her progeny will continue the cycle. If not, there are others who will carry on the life of the water and skies.

Is life any less precious for its brevity? Who will live? Who will die? Some days, it seems like a random roll of the dice. Sigurd Olson, who penned many books about the North Woods and its natural communities, wrote, "Since we go through this strange and beautiful world of ours only once, it seems a pity to lack the sense of delight and enthusiasm that merely being alive should hold." We must stay awake. Pay attention. Keep rekindling our sense of wonder.

"Beauty and grace are performed whether or not we will or sense them," writes Annie Dillard. "The least we can do is try to be there." Perhaps it's enough to show up. To live always on the edge of delight. To be aware of each dragonfly or damselfly that crosses our path, as well as the astonishing joys the natural world holds for those who pay attention. To embrace change, even when it seems negative. To appreciate each day, this day, that we've been given.

To never stop going outside to look. And then, to be grateful for the time we have to do so.

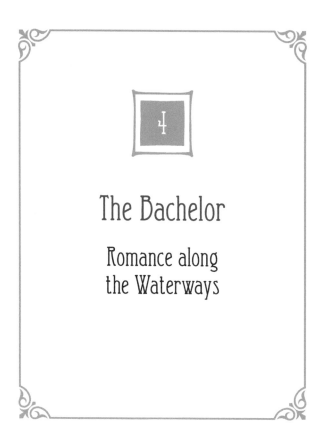

4

The Bachelor

Romance along the Waterways

Many of the reproductive habits of insects are so strange they are almost beyond belief.

—EDWIN WAY TEALE

Dragonflies and damselflies hold important clues to our climate, our aviation design, and the quality of our waterways. But there are more reasons to care than the sensible, practical ones; more reasons that keep dragonfly monitors walking a route. There's the chance to observe—close-up—one of the most bizarre and miraculous reproduction rituals in the animal kingdom.

Dragonfly mating.

It's on display everywhere, wherever I look. In the air. Along the creeks. Deep in the grasses. Walking by a summer stream in the afternoon during dragonfly season is a bit like strolling the French Quarter in New Orleans in the late hours on a steamy evening.

Over there. Nine pairs of Variable Dancer (*Argia fumipennis*) damselflies, each couple locked together, with the male firmly grasping the "neck" of the female. They look like purple pins stuck haphazardly into a watery pincushion. What—are they ovipositing in that decaying mat of vegetation? They are!

And here. Two Ebony Jewelwings in the wheel position (more on that in a moment). He's checking her for another suitor's sperm and . . . found it. Now he's scraping it out. However, it's likely she has her own cards to play. She has sequestered some of her previous boyfriend's sperm where the new guy can't reach it with his special tools. Who will be the daddy? Maybe—both of them! It's an Ancestry.com nightmare waiting to happen.

Once you start looking, dragonfly and damselfly sex is everywhere. Male Odes are staked out at even intervals along the creek, one on every "stream corner." Ready for procreation, they wait for the right female of the species to fly by.

These males are show-offs. Kurt Mead, in *Dragonflies of the North Woods*, tells us males compete for females, with the competitions including "sparring, flight contests, and threat displays of bright colors on the abdomen or wings." Only the dominant males, he says, will get a chance to mate. The rest are the losers, driven away.

How do damselfly males know that the object of their attention is the correct species? In the case of the Ebony Jewelwing, the female's black wings with their bright white dot tell the males that she's the appropriate type. Flight patterns are also believed to be a factor. For dragonflies, the body color also plays a role.

Once the dragonfly male spies an appropriate female, he grabs her by the back of the head with special body parts called *cerci* and a lower *epiproct* at the very end of his abdomen. (Damselfly mating is similar, but slightly different.) When dragons or damsels are in this position, they are said to be "in tandem." The male moves his sperm packet from under abdominal segment nine to the secondary genitalia, located under segments two and three. If the female is interested, she'll bend her abdomen toward him, linking to his secondary genitalia. This forms the "wheel position," which as anyone will tell you, "looks like a heart." The Odes are also connected in two places—a very unusual coupling in the natural world.

The two mating insects may remain locked in the wheel position for as short a time as fifteen seconds or as long as an hour or more, until fertilization is complete. If you are strolling by a pond, you'll see a lot of dragonflies and damselflies in this position, flying up at your unwanted interruption.

Generally it's rough-and-tumble without a lot of romance. However, according to Marla Garrison, author of *Damselflies of Chicagoland*, there is some courtship in the Odonate world. Garrison's colleague Ken Tennessen has shown that contact involves a sensory stimulation of the hairs, or *setae*, of the female by the male of the same species. If for some reason two different species hook up, "the female will refuse to mate because she is not 'stroked,' so-to-speak, and therefore, not enticed," Garrison explained. "Jewelwings are well known for it [courtship]. Some species of Jewelwings actually flutter crazily in front of the female while she perches on overhanging stems above the water, then dive onto the surface of the water and let the current take them downstream to show the female what a good healthy current it is for oviposition."

What if the female isn't interested in that particular male? She might bend her abdomen away from her prospective partner. Females of at least one species of dragonfly in Switzerland go to a much greater extreme, however. As described by Sandhya Sekar in an article for *National Geographic* on a study in *Ecology Journal*, "Called sexual death feigning, this behavior evolved to protect females against aggressive males; for instance, female Moorland Hawker dragonflies risk injury and sometimes death if coerced into mating." When the aggressive male approaches, the female dragonfly freezes, crashes to the ground, and lies motionless. According to Sekar, five species are known for this behavior, including the aforementioned Moorland Hawker (*Aeshna juncea*). However, one dragonfly expert told me he thinks this may be overstated. Females, he said, may just drop out of the male's sightline onto the ground or vegetation, since males are attracted to moving females.

Who knows?

You could spend hours sitting by a stream or pond in July, watching the dragonflies and damselflies mate and then lay their eggs. Females may lay several thousand eggs over their lifetime, in different batches. The different species all have their various methods of ovipositing. Eastern Amberwing (*Perethemis tenera*) dragonfly females fly and tap their abdomens into the water, letting eggs spill out, as males "hover guard" overhead. Ditto Eastern Pondhawk (*Erythemis simplicicollis*) dragonflies. Common Green Darners tend to make a letter *m* in a floating mat of vegetation, with the male in front guarding while still holding the female by the back of her head in a position to lay eggs. The male Variable Dancer damselfly looks like an upside-down music note as he continues to grasp the female, who makes a letter *n* below, laying her eggs in a vegetation mat. Some dragonflies go it alone, with the female ovipositing without the male present. Other dragonflies lay their eggs in mud or just scatter them across a promising area. Laying eggs in dense vegetation, where the female is less likely to be seen, is another strategy to avoid aggressive males. Most eggs will hatch in as little as five days to up to eight weeks, depending on the species, although there are exceptions. Some species' eggs will overwinter, then hatch the following spring.

Dragonfly mating is one of the most bizarre and beautiful couplings in nature and the insect world, and it's happening all day during the dragonfly season in your region. Just imagine what else is going on in the natural world under your nose, right this moment. Whenever I observe these Ode hookups, I wonder: How did I miss this for so many years?

Walking my routes, I muse over the time I spend watching dragonflies. What else would fill those hours if I didn't spend hours walking, looking, and thinking? How did I miss this rhythm of emergence and teneral; flight and migration; mating and new life? Would I be a different person if I didn't pay attention to dragonflies now?

Most definitely, yes. I don't know how to separate myself from the time "squandered" on gazing at insects and baptizing myself in the natural world. Marveling at dragonfly mating habits. In awe over their unusual couplings. Sure, I'm bug-eaten. Mud-splattered. A little sunburned. Yet also: *Relaxed. Calm. Joyful.*

When I'm not out chasing dragonflies during the summer and instead am home, perhaps watching the latest episode of *Chef's Table* or paging through a book, there is the nagging awareness that somewhere, on one of the prairies where I monitor, amazing things are happening. New dragonflies are moving through the prairie. Damselflies are mating and ovipositing in the creeks. Maybe . . . just maybe . . . the Hine's Emerald (*Somatochlora hineana*) dragonfly I've searched for might be patrolling a stream. There's plenty of romance going on along the waterways, no doubt. *And I'm missing it all.*

I can assuage my restlessness for a while by cruising my backyard, scanning the suburban sky for a flier or two, and checking for damsels in my small pond. But usually, after a day or so of this mental tug-of-war, I'll throw my hip waders and dragonfly net into the car, and head off to get a dragonfly fix by walking one of my routes. A little dragonfly romance is a reminder of how marvelous the natural world can be.

I don't want to miss a thing.

5

They Go Where?

The Mysteries of Migration

Not all those who wander are lost.

—J. R. R. TOLKIEN

Dusk. I'm at Kankakee Sands, a Nature Conservancy site in Northwestern Indiana to see the new herd of bison and restored tallgrass prairie with my husband, Jeff. But—no bison at the "bison viewing station." *Ah, well.* We step out of the car to stretch our legs.

"Hey, there are a lot of butterflies over here," Jeff says, gesturing to the prairie, full of bright goldenrod in the dying light. I shield my eyes against the sunset and peer into the tallgrass.

Monarchs! My delight turns to amazement as I look—really look—across the prairie. Hundreds, if not thousands, of orange-and-black butterflies are taking a nectar break on the stiff goldenrod. The prairie grasses and late autumn wildflowers ripple with wings. Monarch migration in process. It's the first time I've ever seen it. We can barely tear ourselves away as the sky grows dark.

Better than bison.

Monarch migration is a phenomenon that is widely known. But did you know that several species of dragonflies also migrate? Some may fly thousands of miles. Trying to understand this mystery is keeping a lot of interested people like you and me up at night. Where do dragonflies go when they head south? What is the benefit to them?

Here in the Midwest, I often see dragonfly migration swarms moving across prairies, interstates, and backyards from late August into September. This seasonal rhythm is surely hundreds of thousands of years old. Yet, it mostly goes unnoticed.

I think of this process one evening in late August. I'm standing on a bridge that spans Willoway Brook at the Morton Arboretum's Schulenberg Prairie. Hundreds of Green Darners are circling me in a vortex of glinting greens and blues. There are other migrants in my area—I look for Black

Saddlebags (*Tramea lacerata*), or the occasional Wandering Glider (*Pantala flavescens*), Spot-winged Glider (*Pantala hymenaea*), or Variegated Meadowhawk (*Sympetrum corruptum*) tossed into the mix. But there seem to be nothing but Green Darners. They zip and dart, a frenzied cloud of motion.

Purple martins swoop in, snatching them on the wing. To a bird, this swarm of dragonflies must appear to be an all-you-can-eat buffet. I marvel at this storm of wings, this maelstrom of insects, all attuned to some wild desire. *Go south! Move!* Why, I wonder? Why these species, and not others? Why today? What convergence of temperature, day length, or genetics is telling them *go Go GO?*

There is so much we don't know. But we do know that dragonflies swarm together for several reasons. They mass together to feed, especially if there is a particularly nice midge hatch on a river. They'll sometimes form a small swarm at the top of a hill. But primarily, in my part of the world certain species of dragonflies in late summer and early fall are coming together to begin their journey south.

We know, Ode expert Kurt Mead tells us, that some Green Darners from the upper Midwest will travel at least as far as the Gulf Coast, Mexico, and the Caribbean. According to Michele Blackburn, an endangered species conservation biologist, each adult will travel about 1,850 miles, a distance which is also similar to that of monarchs. However, unlike monarchs, Green Darners don't cluster in large groups at the end of their journey; rather, they disperse, lay eggs, and die. In spring, the offspring of these migrant Green Darners travel north at about 17 miles a day, writes Blackburn, then reproduce and die. Their offspring head south in the fall, and the cycle continues.

Dragonfly researchers have found some Green Darners also eschew the journey south and, not ready for emergence, overwinter under the ice as nymphs. In the spring, these newly emerged dragonflies mingle with the migratory ones returning from the south. A Green Darner is always the first species I see in the spring, the migratory ones ragged and weary from the long journey. But, as T. S. Eliot lamented, "April is the cruellest month." Sometimes, dragonflies run into trouble. I might see the first Green Darners on a balmy day in mid-April; the next day, the magnolias and daffodils that have just come into bloom are smothered by a spring snowstorm.

I witness this one spring. Only a week before, Green Darners had arrived from the south and were mating in the ponds. These same ponds are iced over today. What has happened to those early arrivals? I don't know. Are the adults perishing in this blustery weather? I have so many questions.

In spring migration, dragonflies tend to return singly, rather than in large groups. This part of the migratory cycle can be unnoticeable, unless you're looking for those lone insects. But fall migration is a stunning mass movement of dragonflies.

In early September, my cell phone dings as I finish dinner. It's a text from our new ecosystem restoration scientist, Dr. Elizabeth Bach, with the Nature Conservancy at Nachusa Grasslands. She oversees the work that my six dragonfly monitors and I do on these 3,800 acres of mixed woodlands, wetlands, and prairies. Elizabeth tells me she's stepped outside the house on the preserve, right before dusk, to find the skies full of dragonflies.

Although she had never monitored dragonflies before, Elizabeth knew that what she was seeing was a big deal. "What should I do?" she asked. I thought for a moment. I was ninety minutes away; the sun would set before I could get to Nachusa.

"Estimate!" I told her. And she did. In her small area, over the course of that evening, she saw hundreds of Green Darners. A stroke of serendipity, always, to catch this event. By mentally sectioning off the sky, she was able to get a general idea of how many dragonflies were in a particular mass.

Her first migration swarm! Like an initiation. So much about dragonfly monitoring is being in the right place, at the right time. And a willingness to pay attention, as Elizabeth does.

Seeing the swarms of dragonflies making their way south is a reminder of the mysteries of migration. Since the 1880s, scientists have been aware of the phenomenon, but it is still poorly understood. There are a few things we do know. We've found that dragonfly migration takes place on every continent except Antarctica. In Illinois, we know at this writing that there are at least five regular migrators: Green Darners, Black Saddlebags, Wandering Gliders, Spot-winged Gliders, and Variegated Meadowhawks. Carolina Saddlebags (*Tramea carolina*) and Red Saddlebags (*Tramea onusta*) are likely migrators as well. The Migratory Dragonfly Partnership estimates that there are as many as eighteen migratory species in North America.

Migrating dragonflies may move with weather fronts, Ode expert Marla Garrison tells me, as wind reduces their energy needs for long-distance travels. They also fly along geographic ridge lines such as mountains or rivers, she said.

"What about damselflies?" you may ask. "Do they migrate, too?" As of this writing, the answer is thought to be no, Dennis Paulson, author of *Dragonflies and Damselflies: A Natural History*, tells me. Only dragonflies. Damselflies may be dispersed some distances, however, carried by wind. But they don't have a migratory cycle that we know of.

Birds are perhaps the best-known migratory species. In

his 1999 book *Living on the Wind*, Scott Weidensaul writes of going to Veracruz, Mexico, to count hawks in migration. As he counted, he was surprised by what else he saw. "Each day, tremendous clouds of Green Darner dragonflies, probably numbering in the millions, would stream by us while we were counting hawks, the dry rattle of their wings sounding like sleet on dead leaves."

Later, writing about bird migration, Weidensaul notes, "That such delicate creatures undertake these epic journeys defies belief." Bird migration makes the mind boggle. To be so small, and travel so far so quickly! Yet how much more amazing is it that a dragonfly, no bigger than my index finger, does the same?

Some dragonflies may travel over oceans—from California to Hawaii (2,400 miles) or even farther, observes Paulson. Studies, including one by Dr. Jessica Ware at Rutgers University–Newark, estimate that the Wandering Glider may cover up to 5,000 miles in migration—double the longest migration path of any monarch.

How do we know the dragonfly migration routes and distances? The Migratory Dragonfly Partnership, which represents a range of organizations concerned with Odonates, is working to engage members of the general public with dragonfly migration through regular monitoring and centralized reporting. Unlike tagging monarch butterflies, studying dragonflies by marking and recapturing them has proved impractical. Green Darners were once studied by using tiny transmitters, but this has also proved unworkable.

However, dragonfly wings and also dragonfly exuviae, Blackburn says, have "stable isotopes" characteristic of the water where they emerged that allow scientists to determine the latitude of their natal ponds. By looking at these isotopes, she says, researchers learn more about the direction and distance of migrating dragonflies.

The migratory trip is fraught with perils. Migrating American kestrels and merlins take advantage of the "aerial buffet," Mead says. Paulson tells me he's heard of Mississippi kites living off migrating dragonflies. Predatory birds aren't the only issue. Edwin Way Teale writes of the entomologist who walked the beaches of Lake Michigan after a two-day storm and discovered large masses of migrating dragonflies washed up, "50 dragonflies for every three feet of his advance." Drowned, presumably, in the mayhem.

If so many dragonflies are weathering midwestern winters under the ice as nymphs, why go south? What is the evolutionary push? What gives my Green Darners at Nachusa Grasslands and the Arboretum the itch to wander?

We know temperature is said to play a role. But we don't really know all the reasons. We do know, Paulson tells me, that natural selection plays a part, as migration enables dragonflies to have more offspring. And Mead reminds us that some dragonflies migrate both as "a strategy to survive winter and improve their reproduction potential." Dragonflies will also disperse, a movement different than migration. Paulson tells us that some dragonflies in arid regions will disperse or leave natal waters if they are drying out and travel great distances to find new breeding sites.

Citizen science and the work of researchers will help us to continue to learn. And we must learn, because what is at stake is too great to be ignored. Our lives are intertwined with the fate of insects. As the climate changes, migration patterns may also change. What would the repercussions of this be? We don't know.

These are lofty—and practical—reasons to care. But there are other reasons to care about dragonfly migration that are more elusive. I like what Barbara Hurd writes about her passion for wetlands in *Stirring the Mud: On Swamps, Bogs, and the Human Imagination*:

I am . . . following a calling I don't understand.
Something carnal and mythic. An invisible,
relentless calling. *A yes, do it, go.* Year after year,
I get up and come to the swamp. *Day after day,
I sit down to write. Month after month, John Muir
wandered a thousand miles, from Indiana to the Gulf of
Mexico.* . . . Who can explain what propels the
monarchs to cross a continent, what tells the
bamboo in Malaysia and Belize that after a hundred
years, today is the day to bloom? There are millions
of calls, a thousand summons, most likely a daily
herald. We get up and go or we don't. We turn,
sniffing the wind, looking for signs.

There is something about witnessing dragonfly migration that stirs the blood, that reinvigorates my sense of hope about the natural world. When the last dragonfly of the season disappears from the skies, I look south . . . and I wonder. When will I see the first Green Darner next spring?

A fine day in April comes along, and out of the corner of my eye I catch a flash of green and blue. Gleefully, I mark the date on my calendar. I feel a renewed sense of hope about the natural world. Parts of it are broken, seemingly damaged beyond repair. But dragonfly migration, at least, continues.

The more I learn about dragonfly migration, the more I find there is to know. The questions in my mind come thick and fast, like kids on summer vacation running to an ice cream truck. I pore over books, read the latest migration findings. Talk to experts. And still. It's a complex phenomenon. There is so much to discover.

Living with a little mystery can be a beautiful thing.

6

So Many Dragonflies

So Little Time

Diversity creates the biological tensioning that makes life in general vigorous and sustainable. It's diversity that ensures perpetuity. The loss of diversity, on the other hand, threatens all life with extinction.

—BARRY LOPEZ

I once owned a T-shirt that said, "So many books. So little time!" Truth. I love to read. It's an endless pleasure, with new books published each week, waiting to be discovered. Old books–good friends–sit on the shelf, ready for reacquaintance. I can't lose a Robin Kimmerer title and replace it with a Mary Oliver; nor would I swap a Gretel Ehrlich for a Paul Gruchow. Each is irreplaceable. A particular book carves its own niche into the literary world and into our imaginations. These stories have become part of my story.

So it is with Odonates. Dragonflies are an endless pleasure, although I haven't seen "So many dragonflies, so little time!" on a T-shirt. There is enough variety in the Ode world to fascinate even the most jaded naturalist. Each species has its own story and its own place in the imagination. Each becomes part of our personal stories. If you are already passionate about dragonflies, you can probably name a species that is particularly important to you.

If I had to choose favorites, I'd be hard-pressed to do so. That said, consider the Calico Pennant dragonfly. First of all, the name. Super cute. These are dragonflies made miniature. Adults are just a tad over one inch long from face to the tip of the abdomen. The male has a line of tiny red "hearts" on top of his abdominal segments; the female's markings are yellow. They are a favorite of my dragonfly ID classes. Post a photo of one on social media and you'll get *Like! Like! Like!* Who knew a creature like this could be found in nature?

Eastern Amberwings–even smaller than the Calico Pennants–are so delicate, so ethereal, you almost miss them at first glance as they weave through the grasses and dip their abdomens into the ponds to tap out their eggs. The males, with their transparent namesake amber wings, and the females, with their patches of golden brown, are enchanting fairy creatures.

A few of my monitors are fascinated with the Baskettails—some difficult to distinguish from each other until you have them in your net. The Prince Baskettail (*Epitheca princeps*) seems so huge when seen in the field! It doesn't happen often that I see the Prince, so it's a treat when I stumble across it.

I found the Clubtails more of an acquired taste. Perhaps it's because I see fewer of them than I do the Skimmers, which greet me each season as part of the regular rhythm of a monitoring route. Compared with the delicate Pennants, the Clubtails look buff and muscular and ready to take on the world. Big, thuggish dragonflies. Face off with one and you'll be reminded of the predacious nature of the Odonate. I don't see a lot of Clubtails on my routes, so discovering one is a delight.

Rarity has a role in playing favorites; the joy of discovering something new. When I found the Red Damsel (*Amphiagrion*) damselfly for the first time, I was thrilled because I had read that it was rare or uncommon in my state of Illinois. Same with the Springwater Dancer (*Argia plana*) damselfly, which didn't appear in my local dragonfly ID guides at the time. I had to have it confirmed by an entomologist. These serendipities keep me out there, wondering what might appear next.

When you chase dragonflies, you also begin to suspect that different species have different personalities. Any monitor who has watched the Meadowhawks knows a feeling of connection. The White-faced Meadowhawk (*Sympetrum obtrusum*) twists its head toward you as if to say, *What's up?* It gives you a once-over, then hovers, seemingly as curious about you as you are about it. The Meadowhawks seem more gregarious—dare I say extroverted?—than some of the other dragonflies, like the Darners, which won't stop patrolling to give you the time of day. Even the Familiar Bluet's scientific name, *Enallagma civile*, is a tip-off that it's a "friendly" sort of dragonfly (*civile* means "civil"—courteous, polite).

Each species also has its own signature moves. The Familiar Bluet seems to hover low, like a slenderized blimp. It floats from grass blade to grass blade. The Halloween Pennant flutters. Shadow Darners (*Aeshna umbrosa*) move with military precision, patrolling the creeks. Ebony Jewelwings butterfly their way out from reed canary grass along a stream to snag a small insect and then loop back to their starting point.

"I love the names," Cody Considine, the deputy director at Nachusa Grasslands, told me. He had just snapped a serendipitous photo of a Cobra Clubtail (*Gomphurus vastus*) while out mowing. It was the first species record for Nachusa Grasslands; a species seen only because Cody was paying attention.

Those names. I feel like Cody does about them. Prince Baskettail. Shadow Darner. Variable Dancer. Unicorn Clubtail (*Arigomphus villosipes*). What would it mean to lose such creatures in the world? Would we notice?

At this writing, there are an estimated 3,109 dragonfly species and 3,212 damselfly species known to us worldwide. That's just about 6,300 Odes to keep track of. It's not a lot, compared with insects like butterflies, which are thought to have about 20,000 species. Even birds, which are estimated at just under 10,000 species worldwide, have these Ode numbers beat. But there are plenty of dragonfly and damselfly species to wrap our minds around.

Those numbers are moving targets as there are many dragonfly and damselfly species still to be discovered. Odonata expert Dennis Paulson details recent reports of new species from Africa and the tropics, then adds, "Our knowledge of the diversity of the group is increasing so rapidly that it is obvious we still have much to learn!" Think of South America, where new insects are found, or Costa Rica, where an undiscovered Ode species, *Gynacantha varanasi*, was reported in 2019, or tropical islands that are largely unexplored. We can expect to see this number fluctuate as new species are found, species are reclassified, and (sadly) other species become extinct.

So what does this mean for those of us in the Midwest? Get ready for some numbers. In North America, we have around 469 Ode species to learn. If you live in Illinois, the numbers lessen considerably. Illinois has just under 150 distinct species of dragonflies and damselflies. Warmer states, such as Florida and Texas, will have dragonflies and damselflies we won't see up here in the Snowbelt. Even southern Illinois will have a few different species than we have in northeastern Illinois, so the number of species we need to be familiar with is less than the state's total list. For citizen scientists and researchers, knowing the baselines for what is normally seen in a region by monitoring these species and their populations helps alert us to changes and understand if certain species are in decline or increasing.

There are many species in the natural world. Some of them are rarer than others. On the tallgrass prairie, I've been working to save our pasque flower population. Pasque flowers are tiny early-blooming prairie wildflowers, furred and pale purple. Last season, we found only a few tenuous plants hanging on, a sad state of affairs. It's a touchy species, with poor germination and a reputation of being difficult to grow to maturity from seed. After collecting about seventy-five seeds from the mother plant, I direct-sowed a batch into the ground, then our Morton Arboretum greenhouse germinated five seedlings over the winter. This spring, after our prescribed burn, I found more than half a dozen seedlings up and growing.

On this particular one-hundred-acre prairie, we have almost four hundred species of plants. But to lose even one species—this tiny pasque flower—would lessen the delight and diversity of the prairie. Sure, most people would never notice. And yet . . .

I'm reminded of this as I chase Odes. Each species has its own particular glory. Even the strange-looking ones, like Midland Clubtails (*Gomphurus fraternus*), and the drabber ones, like Springwater Dancer females, would leave a vacancy if they were to disappear from my monitoring sites. I search them out and then feel a thrill of pleasure when they appear, right on schedule. Other times I feel a ripple of unease. Why don't I see American Rubyspot (*Hetaerina americana*) damselflies this season? What happened to the River Bluet (*Enallagma anna*) damselfly I saw last summer in the creek? Each species is precious.

When I show the image of a very common dragonfly, the Black Saddlebags—with its gold markings and deep black "saddles"—to my students, they often gasp. "How beautiful," I hear them murmur. Yes! They are. And no less pretty for their ubiquity in the natural world. A loss of any dragonfly species, when measured in the delight even common ones give us, is incalculable for future generations.

The writer Paul Gruchow once observed, "Curiosity, imagination, inventiveness expand with use, like muscles, and atrophy with neglect." I think of this as I sit on my back porch recovering from cancer surgery, watching dragonflies and damselflies. So far, I've counted the Common Green Darner, the Twelve-spotted Skimmer (*Libellula pulchella*), and the Marsh Bluet (*Enallagma ebrium*) floating through the grass by the pond. I'm watching for the Red Saddlebags to show up on schedule, resting on my tomato cages in late August. And what is that zipping over the pond? I can't quite make the ID. So much to learn—just in my backyard.

The dragonflies and damselflies help me exercise my imagination. They keep my sense of curiosity sharp and alive. Each one expands my sense of beauty, and what is possible.

Dragonflies make the world a brighter place.

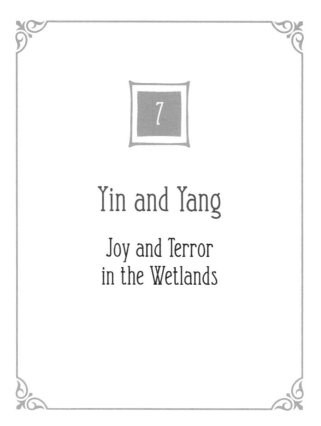

7

Yin and Yang

Joy and Terror
in the Wetlands

Dragonflies were as common as sunlight
hovering in their own days
backward forward and sideways
as though they were memory
now there are grown-ups hurrying
who never saw one
and do not know what they
are not seeing
the veins in a dragonfly's wings
were made of light
the veins in the leaves knew them
and the flowing rivers
the dragonflies came out of the color of water
knowing their own way
when we appeared in their eyes
we were strangers
they took their light with them when they went
there will be no one to remember us

—W. S. MERWIN

A dickcissel sings its signature song under bright sunshine. The smell of something sweet wafts through the air. It's a beautiful day to walk a dragonfly monitoring route at Jay Meiners Wetlands at Nachusa Grasslands.

Ouch! A wild blackberry bramble rips through my hiking pants. Draws blood, I realize, as the drops stain through my khakis. I blot the blood with my bandanna and continue on with more caution. It's a while before I can extricate myself from this thorny patch in the middle of the prairie.

Despite the throbbing pain of my scratches, I feel at peace. Sweeps of coreopsis run like a golden creek across the tallgrass. Bees buzz. Common milkweed is in bloom, the baseball-size flower globes pink and fragrant under the hot sun. There are few noticeable trails. I welcome the chance to move slowly, carefully through the prairie.

One of the gifts of dragonfly monitoring is a chance to look busy without being busy. Walk through a prairie with a camera slung around your neck and a clipboard and you'll appear purposeful. You *are* purposeful. I can satisfy my inner Protestant work ethic, instilled from birth, but still relax and breathe in the smell of a hundred different flowers in bloom across the preserve. Hiking slowly like this helps fill my inner well of creativity that has been drained by the demands of job, family, and even the drive to this place, which involves an hour and a half each way on a busy interstate.

The wetlands here are also a bustling interstate of sorts for insects. It's here that one of my monitors found a new species of Bluet damselflies for our location: the Skimming Bluet (*Enallagma geminatum*). A thrilling moment! It sounds easy when I write this to think of her finding a new species for our site, but looking for damselflies in Jay Meiners is an exercise in blood, sweat, and . . . well, maybe not tears, but the occasional frustration.

It's all about paying attention. Waiting. Seeing what shows up.

Today, what shows up isn't much. I'm logging mostly dragonflies on my data sheet, and not many dragonflies at that; only a few damselflies. Is it too hot? I'm not sure.

And then—there! A Common Green Darner flies out of the grasses and over my head. I start to make a hash mark on my data sheet next to *Anax junius*, its scientific name. The Green Darner is usually our most common dragonfly at Nachusa Grasslands; I'll see hundreds of them this season. But they are no less beautiful for their familiarity.

As I move my clipboard into position, a bird darts in. Snatch! The dragonfly disappears into its beak. Crunch. Munch. Gone.

It happens so fast, my brain doesn't catch up for a moment. *Now you see it, now you don't.* My pencil hovers over the clipboard. Count it? Or not?

As I ponder this question, other questions come to mind. For example, how could a dragonfly make it all the way from egg to nymph to winged denizen of the air, only to be eaten on a beautiful summer's day? It seems so random. A cruel twist of fate . . . although perhaps not for the bird, which got its lunch. Of course, this isn't just a question about dragonflies. As I make the hash mark—it still counts, I decide—I consider the ephemeral nature of this place.

"Nature is a spendthrift only of what she has the most," John Burroughs, a natural history writer from one hundred years ago, asserted. He was talking about plants, but we could apply his remark to insects and, by association, to dragonflies. Insects may make up to 80 percent of the world's species. Recent estimates say we have from two to thirty million species of insects in the world. So what's the big deal about one dragonfly, more or less?

That afternoon, it felt personal. As I've marked my life in rhythms of dragonfly time for the past dozen years or so, seeing their short lives in the air and the precarious vulnerability they exhibit over time, it reminds me of my own mortality and that of the ones closest to me. My baby boomer friends and I are becoming preoccupied with health and longevity. Our bodies are breaking down. A knee replaced here, a hip there. Cancer. Bifocals prescribed and cataracts removed. Moles and suspicious-looking sunspots from a life lived outdoors taken off. A bout of pneumonia. Prescriptions for blood pressure, antacids, high cholesterol. Get a group of us together over breakfast and you'll see—and hear at length—all the indignities of getting older.

Many of us are also in the sandwich generation. We're aware of our comparative youth as we see our parents struggle with late-in-life issues; we're cognizant of our age as we spend an evening caring for our grandchildren while their parents have a night out and then later, after they've gone home, fall exhausted on the couch.

How strange our life is, frost melting on glass, here one moment and gone the next.

These reflections bring me back to dragonflies. Our lives seem short. But the human life span is endless compared with a dragonfly's. Although the nymph may live underwater for as short a time as a few months or as long as seven-plus years—and the average life span of an adult is about a month—the life of a dragonfly in and out of water is a dicey proposition. Consider this: dragonflies, both in the nymph and adult stage, eat one another. So, not only do dragonflies have to worry about frogs, birds, and fish snatching them, they also have to keep one eye on their relatives.

Such a brief and tenuous lifespan! I think of this as I watch a common Twelve-spotted Skimmer circle, then come to rest on a flower spike. He adjusts his wings, long solar collection panels marked with black blotches and blue patches that give him his name. *Libellula pulchella* means, from the Latin, "beautiful little dragonfly." And he is. His short life adds color, joy, and delight to the natural world.

A world that is often anything but delightful. If you are a fan of David Attenborough's *Planet Earth*, you've seen bloody banquets among the mammals and the disregard for anything but hunger and sex, the drives that motivate much of nature. In one episode I've watched hundreds of snakes move toward a baby animal—watched it with my grandchildren, in fact, all under age eight—and just wanted to hit the remote and shout, "No! No!" Some days, evolution seems a terrible thing.

On the Schulenberg Prairie where I'm a steward, I keep an eye on a praying mantis egg case, from its creation in the fall to the early days of spring. I can't wait to see the new baby mantises that will emerge. And then one day, I arrive on the prairie and discover that a woodpecker has ravaged the egg case. All that work gone in a few moments. Praying mantises have been said to bring down and consume hummingbirds. Karma, maybe?

Dragonfly nymphs eat small fish; in turn, Edwin Way Teale spins the story of a two-pound trout that, when slit open by an angler, revealed the heads of thirty-five dragonflies in its stomach. Teale also noted that bats sometimes capture dragonflies, presumably out past their bedtime or not well hidden. Sundew plants may capture and digest some of the unwary smaller species.

"Every live thing is a survivor on a kind of extended emergency bivouac," writes Annie Dillard. Watching dragonflies, I feel the truth of her statement. Teale called dragonfly nymphs "bloodthirsty ogres, stalking endlessly for living prey." Why does a dragonfly nymph eat another dragonfly nymph? Why does an adult dragonfly chow down on another adult dragonfly?

As a prairie steward, I know that without fire consuming

everything in its path every few seasons, the prairie and its rich and diverse community of plants will cease to exist. If the prairie did vanish, certain associated insects would cease living as well. Why do we have to destroy something—literally burn it to the ground—to create a place conducive to certain forms of life? Every spring, I walk the just-burned prairie and turn this concept over and over in my mind. Life from destruction. It seems . . . senseless, doesn't it? Or, perhaps, miraculous?

What grander plan could this be a part of? As Dillard writes, "It's a hell of a way to run a railroad." She adds, "The pressure of growth among animals is a kind of terrible hunger. These billions must eat in order to fuel their surge to sexual maturity so that they may pump out more billions of eggs. And what are . . . [they] going to eat . . . but each other? There is a terrible innocence in the benumbed world of the lower animals, reducing life there to a universal chomp."

A world fraught with perils writ large and small. A world where one moment you're cruising through a prairie full of wildflowers; the next, wiping blood off your shins. A world of pain. Disappointments. Betrayals.

Cancer feels like one of those betrayals. We live in our bodies for decades, and we believe we know them intimately. Or think we do. I thought I knew my own until out of nowhere—that diagnosis.

And I'm not alone. In 2019, more than sixty-five thousand women will be diagnosed with uterine cancer. It's the fourth most common cancer for women in the United States. For those who are diagnosed in the early stages, as I was, the prognosis is very good. Yet in the United States, it is the sixth most common cause of cancer death in women: more than twelve thousand were projected to die from it in 2019. It was a fluke that I mentioned to my doctor my barely noticeable symptoms. Her rapid response to that disclosure likely saved my life.

One week, I hike the prairie, oblivious, chasing dragonflies. Soon, I'm in surgery. In the month afterward, my body aches as the scars heal, ghostly reminders of how life can change in a blink. For six weeks I'm not allowed to pick up anything as heavy as a jug of milk or to walk my dragonfly routes. Something as simple as cooking a meal exhausts me. My memory seems foggy, unfocused. And anxiety hums in the background of my mind. *Did the surgeon get it all?* Time is the only way I'll know for sure.

"Here is the world. Beautiful and terrible things will happen. Don't be afraid." The theologian Frederick Buechner's quote resonates with me. As a person of a constantly evolving faith in something *other*, I cling to his words. *Don't be afraid.* Easily said. But how difficult living it out.

Sometimes, as I hike my dragonfly routes, I discover a spiderweb with a hapless Spreadwing damselfly caught in the silk, wrapped and ready for dinner. Too late for me to meddle. Dragonflies are terrifying to smaller insects, butterflies, and even other dragonflies and damselflies, dashing about and snatching unsuspecting moths and flies while on the wing. And then, the Ode becomes the main course.

On one morning, walking a route on the Schulenberg Prairie, I find a dragonfly lifeless on the ground. It stops me in my tracks. A common Twelve-spotted Skimmer. Kneeling, knees greening, I pick it up and am astonished at its weightlessness. The beautiful wings, once glorious with their sharply contrasting ebony patches that give it its name, are now drab. Researchers and collectors know that once the life force is gone from the dragonfly, the bright colors soon fade. This little guy is too far gone for a collector. The wings are tattered and torn. Did it die of old age? Or was it battling a bird right up until the end? I'll never know.

The world is full of terror, it seems, for dragonflies and people alike. Do insects feel any fear? I doubt it. But I don't know. After all, we're discovering new things about dragon-flies all the time. There is so much none of us understand.

Dillard says of goldfish and their billions of eggs that "nature loves *the idea* of the individual, if not the individual himself—and the point of a goldfish is pizzazz." I would say the same about dragonflies. There is no better "pizzazz" in the natural world.

Learning to live with ambiguity, learning to love mystery, and learning to accept the world on its own terms are all challenges for me. Focusing on the "pizzazz." Anxiety and fear have a tendency to suck the joy out of every positive thing in your life—if you let them. My challenge has been to tip the scales to joy. The dragonflies have helped me make my peace with loving something I don't completely know or understand. With mystery.

In an age when we have the internet and apps to tell us anything we don't know ("Ask the phone," suggested one of my grandchildren, when we were puzzling over some difficult nature question), it's difficult to accept that for some of the hard questions, there are no answers.

Nature—and chasing dragonflies—have a calming effect. I know I'm not alone. While Ohio dragonfly chaser Kim Smith notes that there are other activities that can have meditative qualities to them (knitting or yoga, for instance), she told me there is something special about being outdoors and immersed in nature study.

"I've found that time spent in natural surroundings lifts my mood and changes my focus from whatever troubles I might be having in my life," Smith said. "As I sit on the edge of a pond watching damselflies and see a frog jump up from the water to eat one of them, I appreciate the difficulties of an insect's life compared to the relative ease of my own. I also think about how all of the natural world is connected, and the interrelationships between water, plants, insects, and birds. It gives me perspective on my own life and reminds me of just how incredible is this planet that we're privileged to live on." For Smith, chasing dragonflies is both spiritual and healing, she says.

For a long time before I chased dragonflies, I found courage and joy in solo backpacking. My girlfriends didn't understand this. Wasn't I terrified to be out alone, in wilderness places? And the answer was, sure, I was occasionally frightened. I said jokingly—but also with some seriousness—that I am always afraid of getting lost. Mostly, I backpack on islands. If I do get lost, I reason, I will eventually run into the shoreline. For the most part, being alone outdoors is one of the best ways I can alleviate my anxiety and fears.

Solo backpacking about eight miles to Feldtmann Lake on Isle Royale in Lake Superior, I remember finding solace in a creek flowing under a precarious bridge, full of River Jewelwing (*Calopteryx aequabilis*) damselflies. Their iridescent blue-green bodies and deep black-patched wings imprinted themselves on my mind. I carried the glory of the colors and flash and dance of motion with me the rest of the day on the trail.

That night, putting up my tent after a punishing hike, I felt every muscle relax. I spent the rest of the evening watching dragonflies catching their bedtime snacks along the shoreline before turning in. The beauty of that moment remains clear, even after half a dozen years.

It was at Feldtmann Lake that I saw my first Ashy Clubtail (*Phanogomphus lividus*), which alighted on my arm as I watched the sunset. I moved the slightly torpid dragonfly to my hand and admired it. Down the shoreline, the rustle of a mama moose, bringing in her baby for a drink at the water's edge, broke the silence. I knew wolves were nearby; I saw their tracks imprinting the trails I had walked that same day. It was as close to "wild" as I would likely ever be in my life.

As the temperature dropped, the dragonfly continued

to rest on my sleeve. When it grew dark, I gently put the Clubtail in a brushy area where it had some cover. I slept deep that night, surrounded by the natural world and all its joys and terrors.

The dragonflies have been a part of my desire to live in joy. And by *joy*, I don't mean the superficial happiness of a greeting card. Joy is possible, even when answers are not. Deep joy, I'm convinced, is full of terror, grief, and the cuts from a thousand wounds. It may involve illness. A job loss. The betrayal of a friend. But these things are often balanced by bright moments. Joy happens when you listen to a child read for the first time or you learn a new language or skill. Joy fills you when you watch a mama moose with her baby drinking at a shoreline. Or when a dragonfly you've not seen before alights on your arm.

Joy is often hard-won. Susan Goldsmith Wooldridge, the author of *Poemcrazy*, defines it in this way:

> Joy as I see it involves embracing life. This can
> include moments of sadness, grief or rage as well
> as happiness, unlike depression—where feelings
> are cut off. . . . Joy isn't the opposite of sorrow, but
> encompasses and transcends sorrow. You know
> you're truly connected with yourself when you're
> experiencing joy.

The brevity of the dragonflies' lives, their glorious short time aloft in the world, their fierce beauty—taking on all comers—gives me a shot of courage and hope. Dragonflies keep me in touch with the outdoors. They lessen my anxiety. Contemplating their short lives prompts me to live every minute of mine to the fullest. They remind me to pay attention. Appreciate the moment.

They are portals to joy.

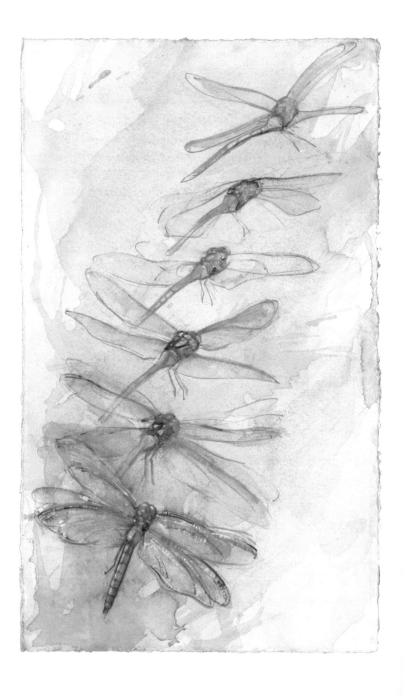

Dragonflies as Creative Muse

Insect-Inspired Art

In human culture is the preservation of wildness.

—WENDELL BERRY

In an old china cabinet, given to me after my grand-mother's death, are my dragonflies. Beautiful pottery mugs with etched wings. Trays with dragonfly images. Bowls covered with dragonfly motifs in blue and white.

There's a large pottery platter with an impressed dragonfly image. "This is mine, but I think this belongs with you," said an older woman, handing it to me at a class. An unexpected gift from someone I hardly knew.

Dragonfly jewelry is in the cupboard as well, scattered among the bowls. Dragonfly earrings from my mother and from friends. A few metal Odonates, some in stained glass, and another in green glass; one dragonfly made of wood.

In my closet are the dragonfly T-shirts and the dragonfly-embroidered blouse. A dragonfly purse hangs on a hook. My cash is in a dragonfly wallet, worked in leather. I have two small dragonfly coin purses, both given by dear friends. In the drawers of my bedroom dresser are pairs of dragonfly socks. On the nightstand is a coaster with the image of . . . you guessed it. I even have an unopened bottle of dragonfly draft IPA beer.

Once you begin chasing dragonflies, your family, friends, and acquaintances will associate them with you. If you're like me, you'll be showered with dragonflies in every possible artistic form. It's astonishing, the generosity of friends and family. Each dragonfly item has an association with the person who gave it to me.

When I pick up a blue bowl covered with dragonflies, I think of my friend Lonnie, who beat cancer. When I drink from the dragonfly mug my sister Sherry gave me, I remember her courage in the face of many difficulties. Mary of the IPA beer was my boss for a while, and now we're good friends. Seeing the beer bottle with its purple dragonfly motif always makes me smile. When I use the dragonfly platter, I think of another friend and the losses she has faced in her life. I'm inspired anew by her love of the natural world.

Other dragonfly items are reminders of the kindness of people I don't know well and yet who gave so freely. It's a good lesson for me in letting go—oddly enough—even as I accumulate more dragonfly paraphernalia.

Dragonflies show up on housewares, clothing, jewelry. *Dragonfly* appears in the name of a Kevin Costner movie and shops and even a water park for kids just down the road. I open my dragonfly books and there are images of all kinds of artistic inspirations, wrought by artisans over time. There's even a Japanese samurai helmet with a dragonfly on top from the seventeenth century.

All of this dragonfly craftwork was inspired by an insect, one most people routinely ignore or dismiss. Yet throughout history, dragonflies have been the inspiration for artists in many forms around the world.

Dragonflies spark music. Do you like something mellow? Michael Franks has a smooth jazz album, *Dragonfly Summer*. More of a rocker? Early in Fleetwood Mac's career—before the addition of some of their main players—the group hoped to hit it big with "The Dragonfly." But the single didn't fly. I asked several of my friends who are hardcore Fleetwood Mac fans what they thought of this particular song. They almost universally answered, "What song?" As soft rock, "The Dragonfly" never really found an audience. But kudos to Fleetwood Mac for trying to raise the profile of Odonata.

The lyrics for the Fleetwood Mac piece, written by Danny Kirwan, came from the writings of William Henry Davies, a little-known Welsh poet. As the story goes, Davies was a drifter who jumped trains in North America until he crushed a foot. He then moved to England. "All the wildness had been taken out of me," he said. But nonetheless he penned twenty collections of poems, some of which, like

"The Dragonfly," live on. In 2011, the musician Blake also recorded a song using the poem's words as lyrics: *It was a fleeting visit all too brief, in three short minutes he had been and gone.*

Other poems are odes to . . . well, Odes. Consider Alfred Lord Tennyson's poem "The Dragon-Fly," which chronicles the emergence of this insect. Or Gerard Manley Hopkins's memorable poetic line, "As kingfishers catch fire, dragonflies draw flame." And a lesser-known poem by Vachel Lindsay from the 1920s, "The Dragon-Fly Guide," compares a "flying machine" and a dragonfly, seen together in the sky, which compete for onlookers' admiration.

Literature in many cultures features dragonflies, especially in Japan. Throughout Japanese history, dragonflies have been featured in haiku and even as a symbol for the island nation. As Forrest Mitchell and James Laswell recount in their fascinating book *A Dazzle of Dragonflies*, one of the old names for Japan is Akitsu-shima, "Island of the Dragonfly." They share numerous Japanese folktales that feature dragonflies, including one in which the dragonfly leads a Japanese couple to a clear spring of wine, or saki (depending on the version you read), and then, great quantities of gold, which make them wealthy.

The seventeenth-century Basho's apt haiku is one of the most lovely descriptions of the dragonfly I've ever read:

Crimson pepper pod
two pairs of wings and look
darting dragonfly

Contemporary dragonfly chasers Scott King and Ken Tennessen present similarly themed poetry in their aptly named collection *Dragonfly Haiku*.

Other forms of writing also reflect a fascination with dragonflies. In North American literature, Mary Webb's *Precious Bane* from 1924 chronicles the story of Prue Sarn, a woman in 1800s Shopshire, England, who has a harelip that draws unwanted attention harmful to her self-esteem. The dragonfly's transition from nymph to adult is used as a symbol of her brother Gideon's wrestling with his dark side—indeed, Prue believes, his very soul: "Maybe you've seen a dragon-fly coming out of its case? It does so wrostle, it does so wrench, you'd think its life ud go from it. I've seen 'em turn somersets like a mountebank in their agony. For get free they mun, and it cosses 'em a pain like the birth pain, very pitiful to see." Webb suffered from Graves' disease, an autoimmune disorder that likely hastened her death at age forty-six. To Webb, the dragonfly may well have been a powerful symbol of transformation, from illness to the hope of good health.

In the culinary arts, dragonflies move from abstract concepts to actual food on a plate. In Bali, you might see someone catching dragonflies on a stick dipped in sap and then taking the captured insects home for dinner to fry, grill, or use in soup. Dennis Paulson's *Dragonflies and Damselflies: A Natural History* shows photos of dragonfly larvae on skewers for sale in a Chinese market. *Dragonflies. It's what's for dinner.* (Or not.)

Some people translate their passion for dragonflies into the visual arts. Illinois artist Karen Johnson (whose email handle is "bugjohnson") told me she is especially fascinated by the complex network of dragonfly wing veins and transparent wings which "reflect the sun in tiny diamond facets." She beautifully captures their natural history in photographs and also in watercolor, acrylic, scratchboard, and sculpted jewelry. Johnson also teaches dragonfly-drawing classes to adult natural history students in the Chicago region. Arthur Pearson, another Chicago-area artist, puts his Ode passion into stained-glass works. And Peggy Macnamara, the Chicago Field Museum's talented artist in residence, painted the beautiful dragonfly and damselfly illustrations in these pages.

Sometimes, the insect inspiration comes secondhand. In his third season as an Illinois monitor at Nachusa Grasslands, Mark Jordan said his wife "caught the bug" after seeing him chase dragonflies and has begun a dragonfly quilt. Think of the pattern and fabric color possibilities!

Most dragonfly chasers eventually pull out their cameras, either for ID purposes or just for recording the sheer glorious brilliance of these insects. Rajat Saksena, who monitors in Delaware County, Ohio, finds photographing dragonflies and damselflies rewarding. "Everything about them is challenging and hence fun! One would call it silly, but I consider it cheating if I net them and then photograph them. I work to establish a relationship," he said. When he discovers a dragonfly or damselfly eating an insect, Saksena will work his way toward it until he's about a foot away, the distance he needs for focus. "I feel almost as if I am interacting with them," he told me. "I feel connected. There is an abstract bond that I can feel in that moment. Perhaps we both are observing each other at that time. I feel that we both are curious about each other and have a mutual respect for one another. It's something I just can't put down in words. It is spiritual."

He's not the only one who feels this way. And, in an era where our insects are under threat, we need all the ambassadors for them we can find. "No better way is there to learn to love nature than to understand Art," wrote Oscar Wilde, adding later, "And the boy who sees the thing of beauty which a bird on the wing becomes when transferred to wood or canvas will probably not throw the customary stone."

Maybe it is that "otherness" of such a small flying insect that keeps us inspired to photograph, paint, sculpt, write about, and celebrate it. And perhaps it is "Art," in all its forms, which will inspire us to protect and raise awareness of dragonflies and damselflies.

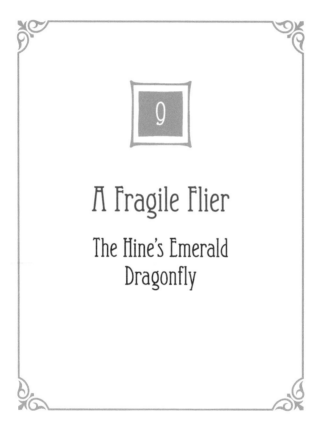

9

A Fragile Flier

The Hine's Emerald Dragonfly

The last word in ignorance is the man who says of an animal or plant, "What good is it?" If the land mechanism as a whole is good, then every part is good, whether we understand it or not. If the biota, in the course of aeons, has built something we like but do not understand, then who but a fool would discard seemingly useless parts? To keep every cog and wheel is the first precaution of intelligent tinkering.

—ALDO LEOPOLD

Some people dream of meeting sports heroes. Others, their favorite rock star. Me, I dream of seeing the Hine's Emerald dragonfly, winging its way through a prairie preserve.

During the winter, I pore over my favorite dragonfly field guides and memorize the markings and natural history of this particular species. But images and words in a field guide are no substitute for seeing the Hine's Emerald in real time.

On this January day, I'm going to get my wish—sort of—at Blackwell Forest Preserve with Forest Preserve of DuPage County ecologist Andrés Ortega. One of his initiatives is raising the Hine's Emerald dragonfly for release in the Chicago region. According to the Illinois State Museum, populations of this dragonfly "seem to be rare and localized" in the state. The Hine's Emerald also is the only dragonfly on the Federal Endangered Species List in the continental United States and Alaska.

I arrive at Blackwell Forest Preserve and drive past the people dropping their lines into Silver Lake. Ice fishing. The Urban Stream Research Center is blanketed with snow. The dead of winter isn't a time most people think about dragonflies. But I know that dragonflies and damselflies are here in the preserve, under the thick layer of ice on the lake and deep in the cold waters of Springbrook Creek. I've seen them in the warmer months, while kayaking. Eastern Pondhawk. Green Darner. Eastern Amberwing. The Hine's Emerald is here also, although not under the ice.

Andrés meets me at the door of the research center and instantly puts me at ease. He began working with the project in the Chicago region in 2017, and his enthusiasm and passion for the dragonfly are evident from the first moment of my arrival. I'm full of questions. Why is this particular dragonfly endangered? What's different about it from its other flying kin?

I soon find out. This is a finicky dragonfly, with special requirements that no doubt have contributed to its demise. The Hine's Emerald lays its eggs in shallow flowing water, preferably in a sedge meadow or fen with a dolomite underlay. Most must spend time during the nymph stage in burrows made by the devil crayfish. Bizarre? Yes! The "devil's darning needle" meets "the devil crayfish." Ironic. Sounds like an advertisement for a World Wide Wrestling match, doesn't it?

How, I wonder, can a dragonfly so particular about its habitat be reared indoors? Ortega reaches into a refrigerator and pulls out a dozen clear vials. Each one is numbered. Each contains a dragonfly nymph. Where did they come from? How did they end up in a refrigerator here? I have so many questions. Andrés patiently tackles them, one by one.

Back in the summer, when the Hine's Emerald was flying in local preserves in Will County, DuPage County, and Cook County in the Chicago region, researchers netted the key specimens: *gravid*—pregnant—females. Such a complex undertaking. Imagine going out to a wetland, looking for a particular species of dragonfly. You have to find a day that isn't rainy, cold, foggy, or windy, as dragonflies aren't always active in those times. The sun has to have been up long enough for the dragonflies to have warmed their flight muscles and shed the morning dew. Yet it can't be so late in the evening that they're already at rest for the night. Dragonflies are *ectothermic*, which simply means they take their body temperature from the temperature around them. Even if it's the ideal time on an ideal day, members of the Emerald family of dragonflies, of which the Hine's Emerald is a part, are notoriously difficult to net.

Once you find the correct species (and remember, this is an endangered species, so the numbers are sharply limited), you have to further parse their number to find females. And not just any females but gravid females.

Then, imagine reaching into the net, and while the gravid Hine's Emerald female is "in hand," the tip of her abdomen is dipped into water. This process, simulating the natural ovipositing process, causes her to instinctively release her eggs. Each Hine's Emerald female lays more than five hundred eggs during her lifetime. The harvested eggs are then driven to a laboratory at the University of South Dakota, where they will hatch and grow in the lab for several years. When the correct time is determined, some of the nymphs are brought back to Illinois and delivered to the Urban Stream Research Center.

On my visit in January, the nymphs were being kept at about forty degrees Fahrenheit in their refrigerator *diapause* (a period of enforced dormancy), which mimics their normal overwintering temperature, Ortega tells me. I look into one of the vials. A nymph glares back. These are ferocious little critters. Don't let their tiny size or immobility in diapause fool you. They are formidable.

As the temperatures warm up outside, the nymphs will be placed into "raceways" inside the stream center. These are long trenches containing running water, sand, frayed rope, and plastic aquatic plants to simulate a "natural" environment. Ortega tells me they keep similar sized dragonfly nymphs together; otherwise, the larger ones will eat the smaller ones. The nymphs will thrive in the carefully calibrated temperatures of the shallow running water in the raceways until they are ready for release. The water, which is piped to the raceways from big barrels, is quality controlled. Not too clean. Not too dirty. Just right.

Why not just preserve habitat for Hine's Emerald? Why go to all this trouble? Less than 1 percent of Hine's Emerald nymphs survive in the wild, Ortega tells me. Pretty slim odds.

Because I use *Dragonflies of the North Woods* as one of my go-to field guides, I page through to the Hine's Emerald description to remind myself what it looks like as an adult. It's a charismatic dragonfly with shiny emerald eyes, one of the key characteristics of the Emerald family. Author Kurt Mead lays out its vital statistics: 2.4 inches long. Mostly yellow face. Adults sport two yellow stripes on their metallic green thorax. A third yellow stripe is on abdomen segment one, with a thin ring around segment two. The male's wings tend toward clear, females are more golden or, as Mead says, "tea stained."

That's enough to remind me of what I'll be chasing this summer. Just for fun, I also check the Illinois Natural History Museum page online, an excellent source of local dragonfly information, where I learn that there is "darker tinting on both pairs of wings as they become older." I can't wait to see the dragonflies in real time.

There aren't a lot of places I can see the Hine's Emerald anymore, although new populations continue to be discovered. The Hine's Emerald dragonfly was first observed in 1931 in Logan County, Ohio. It was at one time seen in Indiana and Alabama, but by the mid-1900s, it was thought to be extinct. In 1983, it was found southwest of Chicago, at Lockport Prairie Nature Preserve in Will County, not too far from where I live. More populations were later found there, and then in other parts of the region. Additional Hine's Emerald sightings and populations were later discovered in Door County, Wisconsin; at least five counties in northeastern Michigan (where it has endangered status and is considered critically imperiled); Ontario, Canada; and the Missouri Ozarks, where it has been found on at least twenty separate sites.

The Hine's Emerald was declared endangered in my home state of Illinois in 1991. It was given federal endangered status in 1995. In Lockport, Illinois, this status led to a bitter battle over drilling wells (which would disrupt the

Hine's Emerald dragonfly habitat) and building an over-pass for a new interstate, involving millions of tax dollars. For dragonfly aficionados, this protection as an endangered species was a victory, but it left other people frustrated. Why go to all this trouble for a "bug"?

For many plant, animal, and insect species, habitat destruction may be one of the biggest threats to recovery. The dragonflies face continued obstacles from contaminated groundwater, road salt, pavement projects, and new construction. Like other flying insects, they also are threatened by vehicular traffic and alterations to local hydrology, such as when wetlands are drained or developed. Droughts, caused by climate change, are also becoming more threatening to Odonate populations.

The U.S. Fish and Wildlife Service tells us that there are between ten million and fifty million species of plants and animals in the world today, and some scientists believe insect species alone are in that range. With these numbers, does one insect more or less really matter? Why all this time, energy, and work for a tiny flying insect? Aren't there more important issues in the world? Shouldn't we let animals and insects go extinct, as part of natural processes? What value do they have to people?

Good questions. And, as Ortega would tell you, the answers vary as to why we invest so much time, money, and energy into fanning the spark of a declining dragonfly species back into flame.

"To keep every cog and wheel is the first precaution of intelligent tinkering," asserted Aldo Leopold, the father of wildlife conservation whose quotation opens this chapter. Leopold noted that when we change the world through human-engineered development, it is our responsibility to ensure that we don't lose creatures and plants in the process. There is a lot we don't know about how the world functions as a community. To lose a member of the natural community may have implications we can't even fathom.

Think of those questions again. The United States Congress answered them in the preamble to the Endangered Species Act of 1973. It recognized "that endangered and threatened species of wildlife and plants are of esthetic, ecological, educational, historical, recreational, and scientific value to the Nation and its people." In this statement, Congress summarized convincing arguments made by scientists, conservationists, and others who were concerned by the disappearance of unique creatures. The U.S. Fish and Wildlife Service also cautions, "No one knows how the extinction of organisms will affect the other members of its ecosystem, but the removal of a single species can set off a chain reaction affecting many others." Wouldn't it be best, then, to hedge our bets and protect and conserve as many species as we can?

The "insect apocalypse" can be seen as a throw-up-your-hands, nothing-I-can-do-about-it situation. Barry Lopez, writing in his book *Horizon* about social justice issues, says, "Distraction and indifference always offer us a way out of dilemmas otherwise too exhausting or harrowing to face." I'd say the same is true for reestablishing dragonfly species such as the Hine's Emerald. Will we make the effort to protect our vanishing insects? Or ignore the problems of extinction and the effort it takes to protect a species?

I think back to the summer I first encountered the federally threatened eastern prairie fringed orchid while I was dragonfly monitoring. I almost stepped on it. Then I saw it, and I fell to my knees. Such a fragile wildflower. Like no other I'd seen before. As I looked around, I saw others, half-buried in the grasses.

I still remember that June day with pleasure and awe. Such a small plant. Such great joy. I imagine what it would be like for my children and grandchildren to live in a world where there is no eastern prairie fringed orchid, no Hine's Emerald dragonfly. Like losing the ash trees that once lined our neighborhood, would we adapt to their absence and move on? Later generations wouldn't know what they'd missed, but their world would be different. Less rich. Less beautiful.

I think of W. S. Merwin's poem "After the Dragonflies," in which he observes "they took their light with them when they went."

As I prepare to leave the Urban Stream Research Center this January day, Ortega tells me that to see the Hine's Emerald adults, I'll need to walk likely areas from late May to early October in Illinois. It's a long shot. But I feel optimistic that this is my year for it. I'm already planning my hike. The Lockport Prairie Nature Preserve today—less than an hour from my home—offers Hine's Emerald dragonfly hikes.

The next time you see a pond or stream, think of the many kinds of dragonflies and damselflies waiting to emerge. And who knows? This might be the year you see the Hine's Emerald, glittering its way through the air.

Me? I'm going to keep looking.

10

The Girl with the Dragon(fly) Tattoo

More Than Insects

Stuff your eyes with wonder.

—RAY BRADBURY

It's the library's annual book fair today. I browse the tables, seeing what's on offer. A face painter waits for customers in the children's area, but at the moment, there are no takers. On impulse, I take a seat. She smiles and shows me her signature art offerings. Flowers. Superheroes. Rainbows. Unicorns.

Of course, I ask for a dragonfly. "Maybe on my hand," I suggest, feeling a bit sheepish. I leave with a stack of books and my dragonfly "tattoo."

My dragonfly "tattoo" only lasted until I washed my hands. Lately, I've been considering a real tattoo—something tasteful that doesn't necessarily have to be in a highly visible location. The appeal comes in part from talking to people who have dragonfly tattoos and hearing their stories. Why are so many people walking around with dragonfly tattoos? Maybe it's because dragonflies are powerful symbols. Perhaps it's their natural history cycle; that nymph clambering around in darkness underwater, then emerging, shedding the old self, sprouting wings, and taking flight.

After I gave a talk on dragonflies and damselflies, a woman patiently stood in line to speak to me. "I want to tell you my story," she said. I hear this a lot—and love it. "My best friend died a few years ago," she continued. "On the day she died, a dragonfly showed up and hovered nearby. Each year, on the day of her death, I see a dragonfly." She showed me her tattoo. I'm not a "woo-woo" kind of person, but I confess, her story—and the seriousness in which she told it to me—made the hair stand up on my arms.

After all, some cultures believe dragonflies are messengers from another world. Who knows?

As I walked into a park district headquarters to speak for a garden club, the middle-aged man working the desk joshed me about my green net. "You giving a talk on butterflies?" he asked. I told him I was doing a program on dragonflies.

Immediately, he walked out from behind his desk. "I have to tell you my story," he said. He recounted how, on the day his teenage son left for college, a cloud of dragonflies appeared in his yard. "Is this a good thing?" he asked seriously.

I told him a bit about the mysteries of migration; dragonflies swarming in late August in the Chicago region are usually ready to head south. "But it's a good omen, isn't it?" he insisted. "Yes, I think it is," I told him. The parallels of a son headed out on his own and the dragonflies readying themselves for their long journey south were easy to see. He seemed satisfied, and perhaps he was comforted by the information.

A father, torn between pride and fear, watching his teenage son leave home for the first time to try his wings in the outside world. A woman, grieving the loss of her dear friend and looking for a sign that her friend's spirit was not eternally extinguished. The dragonflies were there with them, on those days. Witnesses. A part of their lives. A comfort.

Ralph Waldo Emerson tells us that "all natural objects make a kindred impression, when the mind is open to their influence." In Michael Pollan's book *Second Nature*, he notes, "In our eyes spring becomes youth, trees truths, and even the humble ant becomes a big-hearted soldier." He continues, "And certainly when we look at roses and see aristocrats (i.e., "Madame Hardy," "Jacques Cartier"), old ladies, and girl scouts, or symbols of love and purity, we are projecting human categories on them, saddling them with the burden of our metaphors."

I find myself saddling nature with metaphors all the time. A burned prairie becomes a symbol of desolation. A single pasque flower blooming after the burn becomes a sign of hope, of restoration. For someone looking at dragonflies, their journey of incomplete metamorphosis becomes a symbol of rebirth—whether after loss, depression, divorce, or life change.

People of other cultures do the same. Italians reference dragonflies as guardians of the water, or keepers of the fish—which seem like reasonable leaps of imagination, as dragonflies have strong associations with water. One dragonfly monitor told me when she went fishing as a child and the dragons and damsels alighted on her fishing pole, she felt as though they were there to guard her family from flies and mosquitoes.

In many cultures, dragonflies have supernatural powers which they use for good. In his book *The Boy Who Made Dragonfly*, Tony Hillerman retells the Zuni myth from the Pueblo people of New Mexico about a pair of siblings who are accidently left behind when their parents and other adults leave the village to search for food. The young boy makes a dragonfly out of corn as a toy to comfort his sister. The toy comes to life and becomes a liaison between the abandoned children and the gods.

Lovely! But not all dragonfly stories are as touching—or as upbeat—as these. While you'd be hard-pressed to find a negative story about a butterfly, there are many cultural myths that convey frightening images of dragonflies or associate them with supernatural devilry.

Talk to a few people about dragonflies and you'll see how a tiny insect carries the burden of our fears and superstitions. "They'll sting you. Won't they?" These questions come from well-educated people in the Chicago region. The answer is no. Dragonflies don't have stingers, although they do have strong mandibles, or jaws, and—yes, the big ones can nip you if you handle them. But it's very unlikely.

It's one of the first questions my adult dragonfly ID students ask. The "stinger" association is a common one across the Midwest. "When I was growing up, some of the older kids in the neighborhood brought me to a nearby pond where lots of dragonflies and damselflies were ovipositing," Curt Oien, a wildlife technician and dragonfly expert, told me. "My well-intentioned but ill-informed mentors told me to be careful because the dragonflies were 'dipping their stingers.' For years, I avoided them and considered myself lucky that I had never been stung by one."

Despite this early avoidance, Oien went on to become a naturalist and a member of the board of directors of the Minnesota Dragonfly Society. He's a passionate champion of dragonflies. And he has a chance to correct a prevalent myth.

In his wonderful book *A Dazzle of Dragonflies*, coauthor James Lasswell notes, "My grandmother was absolutely sure that 'devil's darning needles' were poisonous (they are not) and often admonished my brother and me for trying to catch one. She told us that if they stung us we would be sick for a long time and might even die."

Those sharp-looking abdomens! It's not hard to see where the fear of being stung comes from. That alone has kept people from investigating dragonflies. If that's not enough, consider how Satan figures prominently in dragonfly lore. The French called dragonflies "the devil's agent," "the devil's needle," and "the devil's hammer." In Spain and Romania, the dragonfly was nicknamed "the devil's horse." One Swedish nickname for dragonflies is "the devil's steelyard"—which was a way of saying that the dragonfly could weigh your soul. Germans referred to dragonflies as "the devil's bride" or "the devil's grandmother." The devil's *grandmother? Really?* One wonders: Why was the dragonfly thought to be symbolic of evil? Does a small insect truly seem so supernaturally frightening?

Much of our fear of insects comes from the unknown. In my work as a naturalist, I find that tiny spiders terrify large numbers of people, perhaps only a bit less than snakes and bats do. As a prairie steward, the two fears I hear from volunteers and visitors who hike the prairie are snakes (especially in the tallgrass, where you can't see them well) and insects. Some of my friends won't hike the prairie with me if the grass touches them—it's not the grass they are worried about, it's the insects on the grass.

I respect this fear. Many people grew up with family members who, upon seeing a spider or ant, acted on instinct and stamped on it or crushed it with a Kleenex. After all, we hire exterminators to rid us of insects in our homes, our yards, and our businesses. A "bug" is almost always seen as something dirty, something pesky, something that bites, or even something sinister. An insect flying around your head needs to be swatted before it hurts you. Or does it?

Dragonflies seem to walk the tightrope between hero and villain; simple "bug" and supernatural god; darkness and light.

Consider this: In Tahitian lore, Teuira Henry tells us, all insects are agents of the gods. The dragonfly was the shadow of a god named Hiro, the god of thieves. Henry explains, "It was a god that flew and halted before and behind. It was carried by thieves in their clothes, so that when they entered the dwelling of those they wished to rob, they let the dragonfly go, and it dazed the inmates so that they did not notice that they were being robbed." Not a good rep for the Odes.

Even the Japanese, who usually attach positive symbolism to the dragonfly, have also given it the moniker "King of Death." Dragonflies and mortality are linked together in the Japanese nickname "Dragonfly of the Ancestors." And in the Philippine Islands, if you tie a dragonfly to your hair, it's said that you'll go crazy.

It's a lot of heavy baggage for a small dragonfly to carry.

When I teach a class on dragonflies and damselflies, I hear other nicknames for them that are less freighted with symbolism and more about their appearance and habits.

References to snakes–"snakefeeder"–or myriad names involving needles, swords, arrows, and pins all make sense because of the shape and body type of a dragonfly. "Mosquito hawk" is another popular nickname in the Midwest, especially from people over sixty: a nod to the mosquito-eating machines that dragonflies are, and their "hawklike" super flying powers.

For others, interest in Odonates is more about facts than folklore. Mention that you monitor dragonflies and they'll pull out their cell phone. "Do you know what kind of dragonfly this is?" they'll ask, scrolling through their photos. I'm delighted they care enough to notice and try to get an image of a dragonfly, no matter how blurry. Or, "This dragonfly fell in my pool and I fished it out," they show me, punching up a picture of a bleached-out dragonfly with ragged wings. "Can you tell me its name?" Chlorinated Odonates can be difficult to ID. But I always try.

One morning, as I write at the coffee shop with a stack of books about dragonflies next to me, I'm interrupted by an older woman. I take out my earbuds and reluctantly give her my full attention. "Do you like dragonflies?" she asks, and adds, not waiting for a response, "I love them!" For ten minutes, she tells me everything she knows about Odes. The facts. But the facts are spectacular, and she takes great delight in recounting them. I nod and listen like it is all new information.

My reluctance to be interrupted by people usually turns to a grudging joy. Wouldn't it be great if more people were like this woman, who had so much passion for dragonflies that she'd introduce herself to a stranger, just to talk about them?

For those who delve into dragonflies beyond the facts, it's a bit more ambiguous. *Who do you want me to be?* the insects seem to ask us. These myths, stories, supernatural ideals, fears, and fascinations are a big load for a little insect to carry–yet, we continue to reference dragonflies according to our own stories, experiences, and life transitions. You don't usually find the same type of associations about wasps or beetles.

In dragonfly lore, there's a lot of supernatural talk. There's the tradition of a good myth or story. But often, when I hear a story from someone I don't know, it involves loss of some kind. The death of a family member. A job that ended. Divorce. A child leaving home. Knowing how brief and tenuous a life dragonflies have—and the transitions they go through from egg to adult—makes their connection to loss even stronger.

A few years ago, I went through a dark period in my life. I found endless comfort in paddling my kayak along a creek each evening and watching River Jewelwing damselflies. I knew the damselflies were insects, understood much about their natural and cultural history, had seen them in their different stages of transformation, and had held different dragons and damsels in my hands. With my rational brain, I looked at them as members of the wetland community and appreciated them as such.

But there was solace, somehow, in their looping flights across the stream, their iridescent bodies throwing emerald and sapphire glints, their total disinterest in me and my despair. Here, floating among the River Jewelwings, there was comfort. This day—this difficult season—would end. The stream would still flow. Dragonflies and damselflies would patrol these waters, adding their bits of brightness and grace to the world. For a short time, I could be a part of that, even if the rest of my world seemed to be falling apart.

Everything has a beginning. Everything has an end. Transitions are a necessary part of a whole and satisfying life. Loss and renewal. Failure and success. The dragonflies and damselflies I saw, as I paddled that stream, reminded me of this.

As they remind me today.

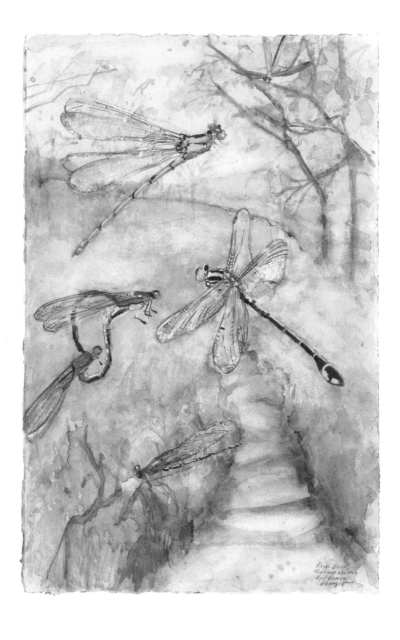

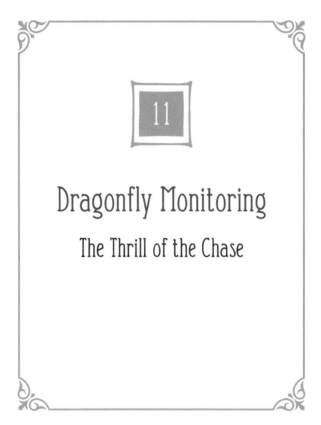

11

Dragonfly Monitoring

The Thrill of the Chase

You can observe a lot by watching.

—YOGI BERRA

The bison facing me weighs over a thousand pounds. And she's not happy. And if mama bison isn't happy, I'm not happy.

I'm walking my dragonfly-monitoring route at Nachusa Grasslands. It's a beautiful circuit, with a large prairie wetland attracting dozens of species of dragonflies. Today, however, it has attracted more than these elegant insects.

Bison once roamed Illinois, not in the huge numbers you see in the old John Wayne cowboy movies, but enough to note their presence. Sometimes called American buffalo, these bison changed the twenty-two million acres of tallgrass prairie in our "prairie state"—from their *wallowing* (rolling around in the dirt), which created minihabitats; to their dung, which fertilized the prairie and attracted insects; to their grazing patterns, which may change the plant composition.

In my early fifties, I became a master's degree student in natural resources; my final project focused on how Nachusa Grasslands could bring visitors to see the bison at this Nature Conservancy site without losing some of the very attributes that made this place so special. In 2014, 30 genetically pure (no cow genes) bison made the trip to our natural area. Today, they've multiplied to a herd of more than 120. I, like so many stewards at Nachusa, feel a delight over their continued well-being.

Normally the bison ignore me on my dragonfly routes. But today I've erred. As I walk over a small slope, unbeknownst to me, mama and her baby are in the pond on the downhill side. Judging by the expression on her face, my puny self is a major threat in this massive mama's mind.

I didn't know I could move so fast. I run to the edge of another pond, smack into a cattail stand. Water up to the top of my boots. Bison, I remember, don't see well. I stand quietly with my clipboard as mama sniffs the air and I try to look like a cattail. *Thank goodness there isn't much wind.* She stares in my direction for what seems like hours, but later, I realize, was only about ten minutes. Then, seemingly satisfied that the threat is gone, she turns and lumbers over the hill.

Dragonfly monitoring is often exciting: seeing new species, listening to birdsong, watching a butterfly float by, and generally enjoying a hike on a beautiful afternoon. But today, perhaps, it is a bit too exciting. The dangers of dragonfly monitoring are usually the typical hazards you encounter outdoors: sunburn, twisting an ankle, mosquitoes, ticks. Bison aren't usually on the list.

Of course, dragonflies face a lot of hazards, too. The peaceful pond, which to me is a place of contemplation and quiet, is a veritable minefield if you are a dragonfly. Underwater, the nymphs face off against turtles, fish, and larger aquatic insects. Above the water it isn't easy, either. A fish waits for the moment a damselfly flies close to the surface, ready to leap and snatch it. Bullfrogs lurk in the shadows. A spider spins her web, entrapping an errant dragonfly. Birds swoop in, catching a dragonfly in flight and leaving only a pair of wings behind. *Snap! Gulp! Gone.*

Such brief lives! Monitoring something so short-lived might seem superfluous to many people. But as it turns out, it matters. Scientists know that in order to understand a species, it's important to track it over many years; even for generations, said entomologist Dave Goulson in an interview with the *New York Times Magazine.* This is not instant gratification. "With so much abundance, it very likely never occurred to most entomologists of the past that their multitudinous subjects might dwindle away," observed the magazine article's author, Brooke Jarvis. "Few thought to measure or record something as boring as their number.

Besides, tracking quantity is slow, tedious and unglamorous work . . . waiting years for your data to be meaningful, grappling with blunt baseline questions . . . and who would pay for it?"

The answer in some cases is this: no one pays for it. Enter the volunteers.

Each season, when my dragonfly monitoring teams and I go out on the prairies and wetlands and woodlands to count dragonflies, we are contributing to something called *citizen science* in the United States. *Citizen science* is simply a term we use for people who aren't academically educated in a particular discipline but who have engaged in some sort of training that allows them to contribute. Citizen science "is often seen by scientists as an idea that successfully combines public engagement and outreach objectives with the scientific objectives of scientists themselves," writes J. Silverton. "A citizen scientist is a volunteer who collects and/or processes data as part of a scientific enquiry. Projects that involve citizen scientists are burgeoning, particularly in ecology and the environmental sciences, although the roots of citizen science go back to the very beginnings of modern science itself."

Without citizen scientists, we would know much less about monarch migration. People like you and me count butterflies in our backyards, tag monarchs, record their tag numbers, and send our hard-won data to collection centers. Then, trained scientists interpret these numbers. Today we know much more about monarch migration because of these efforts.

Although dragonfly monitors don't usually tag the objects of our study, we are alert to changes in the health of our waterways. We watch for species population disruptions.

We chronicle migration patterns. Each of us follows a similar protocol.

Let's take a look.

Every March through October in the Midwest, people like you and me go out and chase dragonflies. Monitors regularly go out to a city park, prairie restoration, pond, local wetland, or other area with the intention of collecting data about Odonates. We spend a good chunk of the summer hours in mosquito-filled areas, counting dragonflies and damselflies and making hash marks on a clipboard to indicate what species we see and how many of each species appear on a certain day.

The Morton Arboretum's one-hundred-acre Schulenberg Prairie in Lisle, Illinois, is where I first noticed dragonflies and began monitoring them. I'm a steward there, so the dragonfly species I see are like the familiar people in my neighborhood. Each season, walking a route is a gateway to meeting old friends I haven't seen for a while, as dragons and damsels emerge on the prairie. It's a heavily trafficked arboretum, with more than one million visitors each year, so my monitoring happens in a highly public place. A special place, and comfortable for its familiarity, like my favorite threadbare afghan or beat-up coffee cup.

Nachusa Grasslands, a Nature Conservancy of Illinois site, is as wild a spot as you might find ninety miles west of Chicago. The western suburbs where I live feature fast-moving cars, impatient salesclerks, air traffic from three airports tracing contrails over my head, and more asphalt than grass. A place like Nachusa Grasslands is my oxygen tank. A visit there helps me catch my breath and slow down. Walking its acres has brought me solace; it's helped me make Illinois my home.

The prairies and mosaic of wetlands and woodlands that make up Nachusa Grasslands are rooted in St. Peter sandstone, and the plants, insects, and mammals that live here reflect that underlay. One of the joys of walking my routes is the carved sandstone cliffs trailing tendrils of vines and plants down their ridged sides. The sandstone *knobs*, or small rounded rocky hills, are what saved the original unplowed prairie remnants from becoming corn and soybean fields. When these precious remnants were conserved, other parcels of land followed: woodlands, creeks, ponds, a fen, and other diverse wetlands. Prime dragonfly habitat.

At Nachusa, I have five regular dragonfly-monitoring routes. A first route contains a wetland that has changed over time from creek to beaver pond to creek again. The dragonfly and damselfly species have changed with it. A second route features a series of large ponds, where I sometimes surprise migrating waterfowl in spring and autumn or stumble across the violet-colored fringed gentian in late summer or early autumn.

A third, one of my favorite dragonfly routes, encompasses a fen and sedge meadow. There are few individual dragonflies but several unusual species for our region. The Midland Clubtail, which is usually found around rivers, turns up here. The Red Damsel is found occasionally on this and other routes. Nachusa also has some common but elegant damselflies, such as the Ebony Jewelwing—the species I first recognized as a damselfly that long-ago day on the bridge, described in chapter 1. The sedge meadow here is home to some of the most vicious biting flies I encounter each summer. I always throw a head net into my backpack for that reason.

A route with a stream comes next, and I'll don chest waders and slowly pull my feet through the sandy bottom, looking for rarities. The River Bluet and Springwater Dancer damselflies are both found around this stream, which is in constant flux as trees along the banks are removed. Rainwater and drought also change the shoreline. The damselflies and dragonflies here reflect those environmental changes from year to year both in species composition and numbers. Part of this particular route takes me to the top of a sandy knob, where I can clock out and sit for a while, enjoying a wide-angle view of the farms to the north and the prairie stretching far below. I put down my binoculars, camera, backpack, clipboard, and camera and let the cool breezes brush away the cobwebs from a busy week at work. The dragonflies fly in clouds around this knob in late summer, so if I stretch out on a hot rock slab and look up, I'll see Black Saddlebags and Common Green Darners flying above my head, silhouetted against the sky.

My fifth regular route is a swampy pond where I've found regal fritillary butterflies and an occasional Band-winged Meadowhawk (*Sympetrum semicinctum*) dragonfly, as well as a Fragile Forktail (*Ischnura posita*) damselfly or two. It's also within the bison area.

More than one hundred bison roam Nachusa Grasslands, and as I said earlier, they occasionally surprise me. One hot summer's day, as I stood ankle deep in the mud by a pond, counting dragonflies, I looked up from my clipboard to see part of the herd silently filing by. They looked like a procession of old women in moth-eaten brown fur coats, moving inexorably toward some distant goal that only they understood. How a group of creatures weighing a thousand or more pounds each can glide so invisibly across a prairie without alerting me to their presence remains a great mystery.

Now it's your turn. Imagine you are a dragonfly monitor. Put on a pair of tall rubber boots. Throw on a mosquito head net. Let's go!

Arriving at your monitoring site, which can be anything from a city park to a wetland to a tallgrass prairie, you step out of the car and hike to the starting point of your route. Sling binoculars or a camera (or both) around your neck, tuck a field guide into your backpack, shoulder a catch net, and pull out your clipboard. You're ready to monitor.

Data gathering begins. On your clipboard is a tally sheet with a list of commonly seen Odes in your area, then a spot to make hash marks as you see individuals of various species, in separate habitats. At the top is your base data. What is your start time? The temperature when you begin? Cloud cover: clear or overcast? Is it still or windy? You'll record the same data when you wrap up your route. This will help scientists understand the conditions under which you observed dragonflies.

Next, you begin walking—"at a steady pace," as the literature advises, but often with a lot of stops to look and get your "dragonfly eyes" on—adjusting your vision to look for small insects instead of the bigger picture of the landscape. You might pause to photograph a damselfly you're unsure of and mark it as an "unknown" on your tally sheet. Or you may stop and admire a Blue Dasher (*Pachydiplax longipennis*) dragonfly, perched on a reed, and then add a hash mark to your data.

The sun is shining, the day is pleasantly warm, and there are few clouds in the sky. Why? Because those are the ideal conditions to go dragonfly monitoring. Most Odes are fairly inactive in drizzle or under overcast skies. They don't fly much when it is chilly. So to walk a dragonfly route is to go for a hike on a gorgeous day. Not a bad deal, is it? To be a monitor at one of my sites is to commit to walking your route at least six times during the flying season. Monitoring is pretty straightforward: Take an hour on a sunny day. Walk your route. Send your data to your site coordinator and then the state database. Done! You've monitored dragonflies.

But I'm oversimplifying a bit, as you've likely gathered from that opening bison story.

Take it from MaLisa Spring, coordinator for the Ohio Dragonfly Survey, a three-year project for outreach, education, and understanding of populations and for obtaining county species lists. Rather than walking specific routes, her monitors are trying to get statewide data, county by county. Spring told me, "Many areas here require permits for off-trail access, so that can take several months. Other areas have poorly developed roads, which can leave you stranded at the base of a washed-out old mining road."

Kim Smith, who also monitors for the Ohio survey, adds that it is challenging to get good photographs of some Ode species for the project: "I also don't have much tolerance for standing out in the hot sun for hours at a time! I've found July and August can be challenging, when temperatures top 90 degrees and you're trying to push through extra-high vegetation from spring rainfalls."

Some dragonfly chasers go to extraordinary lengths to document an unusual species, as did Kurt Mead, author of *Dragonflies of the North Woods*, with three other dragonfly chasers. In July 2018, as part of an environmental grant, they dropped into northwest Minnesota's Big Bog by helicopter to look for the breeding waters of the Quebec Emerald (*Somatochlora brevicincta*) dragonfly. For most of a week, the helicopter, which was too heavy to land in a bog, would hover over the area as the researchers jumped out in ankle-to thigh-deep water.

Mead, who had never been in a helicopter until that trip, said his group stayed in a single spot in the bog for four hours at a time. Then the helicopter would return, fly them to another spot, and drop them into the water again. A lot of trust was required, he told me, as it would be a three-day hike to the nearest road through very challenging terrain if the helicopter failed to return. Using aquatic dip nets and aerial nets, the team would look for flying adults and aquatic nymphs.

Why all this fuss over a single species? It's one of North America's rarest dragonflies, Mead told me, and poorly studied. "No one has described the water chemistry of the pools it breeds in," he explained. Although they saw adult Quebec Emeralds the first three days, Mead said they were unable to find the nymphs, until . . . on the fourth day, Mead recounted, "I picked one up and handed it to Mitch, and he said, *That's it!* I was dancing around! We hit pay dirt!" The thrill of the chase. The rewards of the work.

For most dragonfly chasers who will never see these remote bogs or take a trip in a helicopter to monitor, the rewards are a little less exotic but perhaps no less exciting. Dragonfly monitors care enough to volunteer their hours in the field to accumulate needed data. We sweat, make hash marks on a clipboard, get sunburned, and usually enjoy every minute of it. We keep records for our sites, so we know how our dragonfly populations are changing over time. We upload our data to the state database so the specialists working with it can crunch numbers and make sense of what is happening on a regional, national, and eventually a global level. It's information that will help communities make informed decisions in the future about development, about caring for watersheds, and about responses to climate change.

Mead told me his big push in 2019 would be to survey some of the streams, rivers, and wetlands in northern Minnesota that contain rich copper deposits, which are targeted for copper mining at the time of this writing. He hoped to get a baseline of what is there. Afterwards, it would be easier to see what the impact of the mining is on dragonflies and damselflies. It's this type of important work that lets us understand our human-engineered influence on the natural world. Every species, every individual we count, makes a difference.

Dragonfly monitoring? It's all about enjoying the hike. Tramping through swamps. Even the occasional angry bison. Carrying a net. Getting wet. Sound like fun? For many of us, it is. Making a difference, even in such a small way, is deeply satisfying.

As I write this in early April of 2020, the global pandemic has brought scientific research and Odonate chasing to a screeching halt in the Midwest. The Morton Arboretum has closed. No monitoring can take place on the prairie or in our lakes, streams, and wetlands. At Nachusa Grasslands, we may still hike the trails, but science research and any official activities are prohibited. It's a good call by both institutions, prioritizing the lives of people and their safety over field work.

But for me and my dragonfly chasers at both sites, it's unnerving. For fifteen years, the rhythm of dragonfly monitoring—spotting the first one to return from the south, seeing them emerge from ponds and streams—has shaped my life and the lives of many others. We don't know when life in our region will return to normal. We live with fear and uncertainty.

This viral curveball that has closed off an activity I love has shaken me more than I would have thought. I'm learning more about living with uncertainty and the fragility of human life. As the pandemic unfolds, I take comfort in the dragonflies. They will emerge, unaffected by the troubles we are experiencing. They will continue their life cycle— mating, laying eggs, and then, dying, as they have for millions of years.

The western chorus frogs still call in the wetlands, even when I am sheltering in place at home, unable to hear them. The lavender hepatica wildflowers bloom in the woodlands where I am unable to hike. Sandhill cranes fly over, heading north. In my backyard, I watch spring emerge more closely than I've ever done before. The first chives green up in the garden. The ice melts from the pond, and marsh marigolds open their blooms. In my prairie patch, the rattlesnake master wildflowers push through the soil. All normal. Reassurance, when it is most needed.

The rhythms of the natural world go on. As dragonfly chasers, we take solace in this, even when we are unable to perform our regular observations and data gathering. But I miss walking my routes, clipboard in hand, eyes to the skies.

Dragonfly monitors are all citizen scientists, or, as some people would call them, "amateurs." Jarvis writes that the meaning of *amateur* is "engaging without payment." The root of the word in Latin, she says, is "lover."

It's true. Most citizen scientists do their volunteer work with dragonflies for the love of the outdoors. "If I was going to volunteer my time and energy, I wanted to make a difference in the world," dragonfly chaser Curt Oien told me. "To do that, I would have to do something that everyone else wasn't always doing. I think the dragonfly people are making a difference and that our efforts matter. I keep doing it because there is so much more to be done."

Who are these "Odonataphiles" like Oien who give up their time for free? What else keeps them motivated? Let's take a closer look.

12

The Dragonfly Chasers

Much Ado about Odonates

Hanging out with great and kind people with passions equal to or greater than mine has made my dragonfly work so much more rewarding.

—KURT MEAD

I'm always excited to find a new dragonfly or damselfly species. But it doesn't always happen the way I plan.

On the Schulenberg Prairie at the Morton Arboretum, just outside Chicago, I've been hunting the elusive Red Saddlebags dragonfly. Year after year. No luck. I also look for it at Nachusa Grasslands, ninety miles west of my home, where I monitor several routes. But each season, my efforts are fruitless. None of my monitors have seen this species on these two sites, either.

Almost every year, I find the similar Carolina Saddlebags. When I first get a glimpse of it, high in the air, my heart leaps. Could it be the Red Saddlebags? They are similar. And then I look closer and—no. Sigh. I love the Carolina Saddlebags, but when you've seen that species and not the other, you begrudge it for not being the species you wanted.

One of my passions is gardening. My suburban backyard in a subdivision is small and bordered tightly on all four sides by houses, close to the interchange of highways and interstates, and under the flyways of three airports. Not what you'd think of as a "natural area." But it's a dragonfly-friendly yard, and I'm often visited there by various Odes.

One evening in late summer as I pick tomatoes, I catch a glimpse of a dragonfly resting on a tomato cage. *Must be the Carolina Saddlebags*, I think. Even though I have photos of this species, I can't resist a shot. I get my camera.

The temperature cools as evening approaches. The dragonfly rests quietly under the rampant growth of a Big Boy tomato plant. *Click, click, click*. I reel off a dozen or so photos from several angles and take a moment in the dying light to admire the dragonfly.

That evening, when I pull up the digital photos on my computer screen, something looks different. I can scarce believe my eyes: *Tramea onusta!* The Red Saddlebags—at last. Not in the hundred-acre restored Schulenberg Prairie where I've walked routes since 2005. Not in the thousands of acres of natural areas at Nachusa Grasslands that I've walked the past half dozen years. No. Here, perched on my backyard tomato cage.

It's moments like this that help fan the flame of my passion for dragonflies. Anytime, anywhere—the place and moment when you least expect it—*wham.* A new dragonfly species flies across your path, drops by your garden, or perches on your car antenna. It's those unexpected sightings that keep dragonfly chasers energized after a day of counting the usual suspects (Twelve-spotted Skimmer, Widow Skimmer (*Libellula luctuosa*), Widow Skimmer, Common Whitetail (*Plathemis lydia*), Widow Skimmer, Widow Skimmer, Twelve-spotted Skimmer . . .).

Ask any of my twenty or so monitors who tirelessly dedicate their off-work hours to dragonfly monitoring why they chase dragonflies, and you're likely to get twenty different answers. So much competes for our attention. Work. Family. Screen time. There are many ways to make a difference in our leisure hours, whether it's bagging groceries at the local food pantry or cleaning up litter along a stretch of highway.

Why choose dragonfly monitoring? How do dragonfly experts and monitors "catch the bug"?

For Ode expert Dennis Paulson, it began with a childhood filled with nature and the gift of a bird book on his twelfth birthday. After collecting reptiles and amphibians, he found a work-study job in college that allowed him to focus on birds and vertebrates for a museum. "Sometimes I think about the only group I wasn't interested in by then were the dragonflies," he said in an article for *Argia.* In graduate school, he took an entomology class where he made a collection of Odonates. Fascinated, he changed his PhD work from a study of killfishes to dragonflies. For his post-doc, he reared Odonata nymphs. And, as he said, "the rest is history."

Many have stories like Rajat Saksena, a mechanical engineer doing research in turbulence combustion, who told me his gig as an Ohio dragonfly monitor began with butterflies. As he did butterfly surveys, someone suggested he help out with dragonflies as he was already walking transects. "So, I started noticing the Odes and then just went tumbling down the rabbit hole!" he told me. "I was absolutely mesmerized by these amazing creatures—their color, their flight patterns . . . their eyes are absolutely captivating."

For dragonfly chaser Kim Smith, it began with birds. She enjoyed birding while paddling, and it wasn't long before she noticed damselflies perching on her kayak. "I saw king-birds feeding 'Odes' to their fledglings on the edge of a lake . . . and sometimes I'd find an accidental dragonfly in the background of a bird photo I took." Smith was hooked.

Then, she attended a Dragonfly Society of the Americas (DSA) meeting. It was a turning point. Now, Smith, who is retired, monitors for the Ohio Dragonfly Survey. She also frequently blogs about dragonflies—including her adventures in misidentification—as she engages others with her passion for Odes. Smith said the rewards are in finding a new county record, or getting "a nice, crisp photo."

Ode passion is rewarding—and contagious. Its "conferences" or meetings are a place where experts and amateurs can keep up on the latest in the field, and learn from each other. A DSA meeting is a warm and inviting introduction to the world of dragonflies. Each year, the DSA holds regional and national meetings to bring together beginners and experts from North, South, and Central America. It also publishes two journals: *Argia*, a quarterly newsletter, and the peer-reviewed *Bulletin of American Odonatology*.

Dragonfly chaser Mark Donnelly, who has monitored for many years in the Chicago region, says he and his wife travel to the Wisconsin Dragonfly Society Meeting, the Minnesota Dragonfly Society Meeting, and sometimes the international meetings. "Once I told an old friend from high school, not a biologist, that we were heading to Virginia to attend a Dragonfly Society of the Americas meeting. She replied, 'Sounds like a society where you'd wear costumes and greet each other with a secret handshake.'" Donnelly added, "At least we Ode followers understand each other."

When I attended the 2005 DSA meeting in Decorah, Iowa—as someone who at that time knew almost nothing about dragonflies—members were eager to show me the ropes . . . er, nets. They invited me to socialize, took me out into the field, explained the differences between species, and helped me capture my first Ode, an Autumn Meadowhawk (*Sympetrum vicinum*), or, as I learned it then, Yellow-legged Meadowhawk. I still have that photo, hanging on my refrigerator.

One of the nicest people I met at that conference was Kurt Mead, author of *Dragonflies of the North Woods*. I lined up to have Mead sign my first, spanking-new field guide to dragonflies. In 2019 I reconnected with Mead, an interpretive naturalist at Tettegouche State Park in Silver Bay, Minnesota, by telephone and email and reminded him of that first meeting, fourteen years earlier. He was just as nice as I remembered him then as he told me his story.

Mead said he's been a naturalist-minded person since he was young. As an adult, he looked for a particular place in the natural world where he could become an expert. "I wanted to get proficient in something," he said. Then Mead ran into a researcher doing dragonfly work. "The clouds parted. The angels sang, and dragonflies came flying out of the sky! I was hooked," he said.

When he began chasing dragonflies, Mead said, there was a "large, Minnesota-shaped hole in the distributions of dragonflies and damselflies. We knew so little." It was lonely, he said, and challenging to learn what he needed to know quickly. "Being mostly self-taught, I felt like the research that I had to do on my own was very valuable and rewarding."

Mead started the Minnesota Odonata Survey Project—now the Minnesota Dragonfly Society—and recruited volunteers. Since then, he has helped host everything from the 2018 DSA conference to smaller conferences and workshops. Along the way, he penned *Dragonflies of the North Woods*, a field guide now in its third edition and used by monitors across the Midwest. It remains a favorite of mine.

"The work isn't done. Exciting finds pop up in Minnesota every summer," Mead said. "But getting a 'baseline' of sorts, for Minnesota, has been amazing."

At the 2018 DSA conference, Mead met MaLisa Spring, the state coordinator of the Ohio Dragonfly Survey, when she netted an Eastern Least Clubtail (*Stylogomphus albistylus*) during a conference field trip. Spring remembered Mead's enthusiasm—because he had looked for this species along that stream for years and had yet to find it. "It is nice to see so much excitement from others," Spring told me. "Even when [species] are somewhat 'common' in some places, they can take someone's breath away [in others]."

Spring told me she found her way to dragonflies through bees, which she researched in Marietta, Ohio, as an undergraduate. She later focused on the impact of urban habitat management on beneficial insects like bees, hoverflies, and pirate bugs. As Spring was finishing her master's degree, she heard about a job as state coordinator for the Ohio Dragonfly Survey. "As I already had a wide background in entomology from my undergraduate and graduate work, I was more than willing to jump in and specialize in dragonflies," Spring said. For someone used to 400 to 500 species of bees, 170 species of Ohio Odonates seemed easy in comparison, she said, although "we still have our tricky Odes." Most Odonates can be identified—right down to the species level—from a photo, she added, which is not true of many of the bees.

For Odonataphiles like Spring, finding a new species record at a county or state preserve is motivation to keep monitoring. "I like finding new things, so it is fun to target a new species and successfully find it," Spring says. She found new populations of Lilypad Forktail (*Ischnura kellicotti*) damselflies through targeted searching. "I have also been checking aquatic garden stores and fish hatcheries to find some weird species for Ohio," she noted. At the garden centers, they've seen Rambur's Forktail (*Ischnura ramburii*) damselflies and even a Scarlet Skimmer (*Crocothemis servilia*) dragonfly, which Spring said were likely transported from Florida with plant shipments. Visiting garden centers to look for dragonflies? Anything for a new species! And Spring told me, "It is also really fun to hang out with other dragonfly enthusiasts. We have a nice little community in Ohio, and it really helps to be around the infectious enthusiasm of others to look for Shadowdragons, or Setwings, or whatever else we might come across. The people make a big difference."

✦

Linda Gilbert's passion for dragonflies started when she graduated from college. Today, she's a park naturalist for the Geauga Park District in northeast Ohio and monitors their properties in that county, which includes a kettle lake/bog at Burton Wetlands State Nature Preserve. She's also one of the coauthors of *Dragonflies and Damselflies of Northeast Ohio*, an excellent field guide that my monitors and I use in northern Illinois.

One of the joys of being a monitor, Gilbert told me, is the possibility of discovering something new. Several years ago, Lilypad Forktails started showing up again in various locations in Ohio, she recalled: "One day, I was finishing up some field work and had a few minutes left, so I decided to visit Burton Wetlands to see what was flying. In the back of my mind I was thinking, 'Haha, maybe you'll see a Lilypad Forktail.'" As she looked around, Gilbert saw a small damselfly take off and land on a nearby water lily leaf. "Something about its flight behavior was different and pricked my attention," Gilbert said. "I got the binoculars on it and–holy crap! I recall saying out loud to no one: 'It's a Lilypad Forktail!'" She grabbed her camera and took voucher photos. "It was a county record for Geauga and then I found out it is endangered (in Ohio) too! I think I did a little happy dance on the dock!"

One of the finest dragonfly chasers in the Chicago region is Ode expert Marla Garrison, a biology instructor at McHenry County College in Crystal Lake, Illinois, and author of the field guide *Damselflies of Chicagoland*. "Biology is life–and everything about life fascinates me," Garrison told me in a telephone conversation. Unlike most dragonfly chasers, however, Garrison is enamored of dragonfly nymphs rather than the flying adults. "Very few people are interested in the aquatic stage, and yet, this is where most Odonates spend the majority of their life," she pointed out.

The challenge, Garrison said, is getting good at identifying nymphs. She explains, "Once you start differentiating–and it is a slow learning curve–then the keenest joy I have ever felt in my life is in a flowing stream, pulling out nymphs. . . . You put the nymph under your loupe and confirm it, you know where to find it, the substrate you took it from . . . what its requirements are . . ." Her voice trails off, but you can sense her passion for her work. Chasing dragonflies "has changed my life," she said. You can't talk to Garrison without wanting to plunge into a stream and explore the mysteries under the surface.

For all dragonfly chasers, there is a bit of serendipity involved in the rarest finds. Curt Oien, who retired from Coca-Cola and now serves on the board of the Minnesota Dragonfly Society, told me that he and Mead spent many days near the North Dakota border and along the Canadian border looking for damselflies, with the hopes of finding a species they were sure was in Minnesota at one time: *Ischnura damula*, the Plains Forktail. Oien told me, "Many of us had looked at thousands of specimens over many years, but none of us were able to find *damula!* Were they ever here? Are there any left? Are we too late?"

He told me of grant money spent, trips into remote areas with four-wheel-drive trucks and amphibious vehicles, canoeing, hiking, and hitching rides on Department of Natural Resources helicopters to get to remote areas with rare habitats. No *Ischnura damula*.

A condition of one grant he received was that he or other dragonfly experts would present public programs. Oien told me that about twenty-five people showed up for a program he gave at Norris Camp in the Beltrami Island State Forest. The day was beautiful, he said, and kids were chasing dragonflies through the shallow ponds before everyone had lunch.

One teen, Jason Haag, was a savvy Ode monitor, Oien recounted. "Even so, Jason had one that he wanted me to look at—he said he'd never seen one like it. Instantly I knew. *Damula!*" Oien said, in awe, "A state record species that had been sought for years was found at a public program site with kids running around in a gravel pit that we arrived at in cars!"

As dragonfly monitors will tell you, seeing Odonates is usually a matter of being in the right place at the right time. Showing up. Being present to the moment, as Jason was. Paying attention. Then, being generous with what you learn. Taking time to mentor the next generation of dragonfly chasers–and sometimes, learning from those you least expect.

Not a bad metaphor for life, is it?

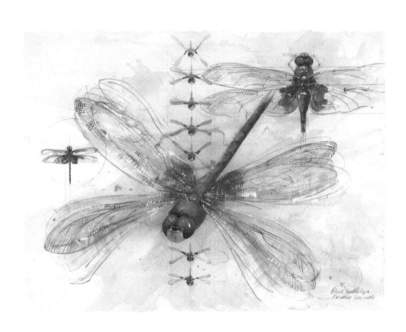

13

Experience Dragonfly Magic for Yourself

Put Your Knowledge to Work

What would the world be, once bereft
Of wet and wildness? Let them be left,
O let them be left, wildness and wet . . .

—GERARD MANLEY HOPKINS

It's the first warm day of April: seventy-four degrees. In the lakes and ponds at the Morton Arboretum, the painted turtles are out. Bloodroot blooms open in the prairie savanna, unfolding from their leaf wrappers. Bees, ants, and flies are zipping around the mostly bare earth.

Common Green Darner dragonflies are returning from the south. Other Green Darners are emerging from the wetlands. Shading my eyes against the sun, I try to distinguish which Green Darners are returning migrants or newly emergent, but I can't always figure out the difference. All I do know is I'm glad—so deeply glad—they're here again.

This season I'm already creating a list of "wants" like a child anticipating Christmas. I *want* to see nymphs pulling themselves out of Willoway Brook, climbing the reed canary grass on the shore, and transforming themselves into colorful fliers. I *want* to discover their exuviae left behind and see if I can learn to identify them. I *want* to find the Hine's Emerald dragonfly at one of my monitoring sites. *I want. I want.* Each year, I have a new list. Each season, I live slightly on the edge of anticipation. Wondering what will turn up next.

You can read about dragonflies, pull up their images online, put on a pair of festive dragonfly-themed socks, drive around with a dragonfly bumper sticker on your car. But at some point, all the head-knowledge and hard goods aren't enough. How do you see dragonflies? How do you get to know them in their natural habitats?

If you live in the Midwest, you'll find the dragonfly season is shorter than you'd think. Birders, who spend 365 days a year glassing the skies with their binoculars, never get a break. Dragonfly chasing in Illinois is confined to the good-weather months. For me, this is a plus: no going out in minus twenty-degree weather to look for likely suspects, as my birder friends do.

For many dragonfly enthusiasts, the "dragonfly experience" begins unintentionally. One of my adult prairie ecology students told me he went for a prairie hike. "There was a dragonfly that had been flying around me for a while when it suddenly zoomed away and headed straight down the trail," he said. "It then turned around and zoomed back, abruptly stopping a few feet in front of me. It hovered for about ten seconds." This encounter made a profound impression on him. There was, he said, "an invisible energy" that seemed to flow between him and the dragonfly. "It felt like we were staring into each other's souls."

Although his encounter was unplanned, you can be intentional when it comes to seeing dragonflies. Looking for a dragonfly or damselfly is usually as easy as finding a retention pond or stream in your neighborhood. Park yourself on the grass for at least half an hour and watch the water. For most of us, who tend to see what we expect to see, it takes a few minutes for our eyes to adjust to looking for dragonflies. Like all exercises in paying attention, it requires a bit of focus, quiet, and concentration.

In her marvelous book *On Looking: Eleven Walks with Expert Eyes*, Alexandra Horowitz reminds us that the world is wildly distracting. Think about the bombardment of sights, smells, and sounds that you are experiencing even as you read this book. Because what we can take in is limited, we have to sort through what we are experiencing and "tune in" to what we specifically want to see, she writes. "Half of tracking is knowing where to look, and the other half is looking," Horowitz says, quoting Susan Morse. Finding Odes takes practice. It takes focus. You learn to selectively see what it is you are looking for. Then, you go outside and *look*.

Damselflies tend to fly low in the grasses. If you sit by a stream, many will be flying around at about eye level. Watch the twigs, the grasses, and the foliage that hangs over the water. Look for damselflies looping out over the surface, hoping to pluck unsuspecting insects from the air, then returning to chow down on them streamside.

If you're sitting by the water, watch for dragonflies flying patrol. They may occasionally splash down, tapping the water lightly, or fly in tandem. You might see them ovipositing in a vegetation mat. Look up to see them silhouetted against the sky, where they may hover for a few seconds, getting a closer look at you and the mosquitoes you are attracting. Then, try to visually follow the dragonfly's flight pattern as it reaches the end of some invisible boundary and returns to patrol the waterway again. Often, if you spook a dragonfly that is at rest and it flies away, it will return in a few seconds to that same perch. No need to go chasing it. It may boomerang back to you.

Different species of dragonflies and damselflies prefer different habitats. A river will have some signature species that you may not find in a pond. And vice versa. You may see Spreadwing damselflies at woodland edges and Meadowhawk dragonflies in prairies, with nary a water source in view. Dragonflies can be found up to three miles away from their natal waters. I've found them on my car antenna while parked outside a garden center. And I've seen them at ballparks, catching insects under the hot stadium lights. Another favorite spot? Traffic lights in busy intersections. You can find dragonflies and damselflies just about anywhere. If you pay attention.

I teach dragonfly and damselfly identification to adults at the Morton Arboretum, and I often have a mix of seasoned monitors and students new to Odonates. We begin with a short PowerPoint presentation on the species they are likely to see that day. Then, we move outside. This past

August, we walked along the prairie trails for our dragon-fly hunt.

You could see the Odes coming into focus for those new to dragonflies. Veterans in the group called them out. "Twelve-spotted Skimmer!" "Common Whitetail!" We waded into the tallgrass, and immediately Calico Pennants and Halloween Pennants fluttered around us, perching on the purple prairie clover and stiff goldenrod plants. "Ohh-hhh! It has tiny red hearts!" exclaimed one woman as she observed the male Calico Pennant through her binoculars. Previously, she said she knew nothing about dragonflies. It was obvious she was now hooked by one of the more charming members of the Odonates.

"What's this?" asked one of the monitors who was in the class for a refresher, stumbling across a Variable Dancer damselfly deep in the grasses. Its vivid color was striking, and this was her first encounter with the species. "They're so much tinier than I thought!" she said, while others nodded. We found masses of Eastern Forktail (*Ischnura verticalis*) damselflies, one of the most common species of damsels in the Chicago region. They are variable in coloration, depending on gender and age. One of my friends, Lily, refers to them as the "Eastern Forklifts"—a wonderful mis-naming that makes me smile now every time I see one.

A pair of "butterfly binoculars" that focus on creatures nearby is helpful when you're chasing dragonflies. Birding binoculars are usually designed to focus on a flier farther away. Ask for "close-focus" binoculars. If you have a camera with a zoom lens, you'll be able to see some of the drag-ons and damsels easily without spooking them. Once you take a solid set of photos, you can enlarge the photos on your computer at your leisure and match your dragonfly or damselfly to an image in a field guide.

As a monitor, I need all the help I can get. I have a mild

form of self-diagnosed *prosopagnosia*, or "face blindness," in which I have trouble remembering people's faces. My husband, Jeff, often pulls off the assist when we are in groups ("Cindy—you remember Rachel . . ."). I believe this affects my ability to recognize dragonflies on the wing, although of course no studies of this correlation have been done. I can ID dragons and damsels much more easily at home, with my field guides and photos, than I can in the field. Knowing this, I always plan to bolster my field observations with images. And, I have great empathy for those who have trouble IDing and remembering species names.

Most monitors who chase dragonflies take photos. Researchers collect voucher specimens, but many of us aren't into acetone and glassine envelopes—although we admire those who carry on the more difficult research. I've found that if I get a good photo of the face, the sides, and the top of the dragonfly, I usually can make the ID without catching an actual specimen.

However, all of this ID work is easier said than done. Dragonflies are quick. Jeff, who sometimes hikes with me as I chase dragonflies, is now an expert at standing in wet foliage and holding plants steady as I photograph a perched dragonfly on a windy day. Chasing dragonflies is an exercise in patience. Sticking with a dragonfly until you can see it clearly. Waiting until the wind dies down for a moment, or the dragonfly shifts its position for the shot. Ignoring the mosquito biting your hand as you try to hold your camera without it wobbling.

As you learn the dragonflies and damselflies at a particular location you frequent, you'll develop a knack for knowing who's who, just by seeing the flight patterns and flashes of color when they are on the wing. You'll build a relationship with the dragonflies in your part of the world. All it takes is an investment of time.

With our increasingly high-powered, lightweight cameras and high-tech binoculars, we have a lot to be grateful for compared to the olden days of dragonfly ID. In one spectacular sentence in his book *The Strange Lives of Familiar Insects*, Edwin Way Teale noted, "Some species are so wild and shy that entomologists have had to shoot them on the wing, using a very fine powder shot to bring them down." Say—*WHAT?* Yes, it's true.

Dennis Paulson, author of *Dragonflies and Damselflies*, told me that while in Costa Rica in 1966, he recalls using a .22 caliber revolver with 125 tiny lead pellets, especially made for collecting dragonflies and other small animals. Using this technique, he was able to capture an undescribed species, the Icarus Darner (*Coryphaeschna apeora*). "Collecting in this way is probably obsolete now," he said. Marla Garrison, author of *Damselflies of Chicagoland*, says the older entomologists she hangs out with tell her stories about putting flour in a gun, which when fired would spray out and weigh down the dragonfly's wings. Then they'd pick the dragonflies up off the ground. These anecdotes make netting them, a somewhat controversial technique here in the Chicago region, seem like a no-brainer.

Most monitors agree that the Odes who perch are easier to ID than the continuous fliers. "The Darners and Spiketails are not easy to photograph or net," says MaLisa Spring, state coordinator of the Ohio Dragonfly Survey. Spring goes to great lengths to get a new species. So does one of her monitors, Jim Lemon, who documented three state records for Ohio: Swift Setwing (*Dythemis velox*) dragonfly, Paiute Dancer (*Argia alberta*) damselfly, and Jade Clubtail (*Arigomphus submedianus*) dragonfly. "Nothing focuses the attention like finding something new," he told me. "For me, this is as close as I come to touching the infinite."

Every spring, the challenge for me is to get my "insect eyes" on. I lose the art of seeing tiny flying creatures in motion over the winter. To begin monitoring requires refocusing my attention and recalibrating my brain, which is accustomed to the more monochrome fields of snow in January and February. In dragonfly season, there is a lot competing for your attention. Wildflowers. Trees leafing out. Butterflies. Birds.

Rajat Saksena, who surveyed butterflies before getting interested in Odes, told me he didn't notice dragonflies at first. "The one thing that completely blows my mind is that before I started contributing to the Ohio Dragonfly Survey, I was already out in the field chasing after the butterflies. All those dragonflies and damselflies that I am now so fascinated with were all there all that time; I just never cared to notice. What a shame! Now I see them everywhere. And that's where I feel pity for the rest of the world. This absolutely amazing beauty of nature is literally right under their noses, and yet they continue to miss it!"

One of the downsides (or is it an upside?) of dragonfly monitoring is you begin to care—deeply—about things you may not have thought about before. A wetland is converted to a shopping center. The new parking lot that makes it easier to shop comes with a price in nature. Instead of celebrating a new coffee place, you wonder which dragonflies—and what associated plants and insects and members of the aquatic community—have vanished to make way for it. You see a meandering creek turn into a tightly channeled stream and you think of the effects on the underlying substrate, the bottom of the waterway. "Habitat loss is heartbreaking to anyone with sentiment for nature," says Lemon.

It's like learning that purple loosestrife is an invasive plant in our waterways. Suddenly, instead of *oohing* and *aahing* over the beautiful carpet of amethyst in the river, you know it for what it is: a take-over agent. And you become *that* person . . . you know the one . . . who points out the problem of the plant to the people admiring it.

But back to our Odes. Learning to see dragonflies comes with a load of add-on benefits. You are aware of shifts of weather. You notice changes in the clouds—from cumulonimbus to cirrus. A sudden temperature drop occurs and you reach for a jacket and also wonder what is happening in the life of the insect world.

When you hike, you swivel your head back and forth, looking down for Familiar Bluets, overhead for Green Darners. You discover layers of paying attention that you carry with you into all other parts of your life. As you photograph the dragonflies, you see the difference of brilliant overhead sun at noon and slants of light at sunrise and sunset. You notice the way a dragonfly holds its wings on a chilly morning and the headstands it does in the "obelisk position" on a hot July afternoon. The pleasing shape of a Blue Dasher dragonfly on a sunflower gives you a jolt of happiness. The glory of teneral American Rubyspots emerging in a flush of invasive reed canary grass are almost enough—almost—to overcome your frustration as a prairie steward managing that same grass.

And that's just above the water. Another one of the joys of dragonflies is that they remind you of the underwater world—that stream, the ditch full of rainwater over there, a pond—and all the life that it contains. You're humbled by how much there is in the natural world that you're not aware of.

On an old world globe, the areas not yet explored were inscribed—*Here be dragons*. Now, every "blank spot" in a wetland or natural area is a possible place for real dragons—those adult dragonflies and nymphs you know are everywhere, but often unseen or unnoticed.

It takes a while to discard other images in your field of vision—trees, prairie grasses, sky—and focus on that glint and dance of motion. But once dragonflies are on your radar, you'll be preprogrammed to "see" these flying insects.

This heightened awareness comes with its own caveat. Suddenly, you realize dragonflies are all around you. And you want to know their names. Wherever you see them. Parking lots. Your backyard. While stopped at an intersection.

For some people, "distracted driving" means texting and talking on cell phones. Those honking trucks behind my car? They're impatient for me to get moving at the traffic light instead of puzzling over the ID of a dragonfly hovering just in front of my windshield.

They could be in for a long wait.

14

Fostering the Dragonfly "Bug" in Kids

Share the Love

*Most children have a bug period,
and I never grew out of mine.*

—EDWARD O. WILSON

Nonna, I caught one!" This from three-year-old Tony. He and I were chasing dragonflies on the Schulenberg Prairie where I'm a dragonfly monitor and steward. In his small net was an Eastern Forktail damselfly. The expression on his face! Wonder. Awe. Surprise. I'll never forget it. I don't think he'll forget that moment, either.

The best way to communicate the importance of dragonflies and damselflies to children is to model that passion yourself. When my grandkids turn three years old, they get their own pocket-sized copy of the Stokes group's *Beginner's Guide to Dragonflies* along with a small butterfly net or, if they are on sale at the hardware store's dollar table, a small pool skimmer net. I try to take them and their nets and field guides to a local watering hole where they can observe dragonflies and make their first identifications from matching the dragonfly they see to the photo in the book (with lots of help from Nonna). Five of the kids have already gotten their books as of this writing; one more to go.

It's not enough to give children information. You have to help give them an experience. On one outing with two of my grandkids, Ellie and Jack, we were all lying bellydown on the boardwalk at the Children's Garden of the Morton Arboretum's Wonder Pond. It gave us a frog's-eye view of the dragonflies flying around, just above the surface of the water. Ellie, then six years old, and Jack, then four, were able to match a Dot-tailed Whiteface (*Leucorrhinia intacta*) dragonfly to the photos in their field guides. We rarely see this supposedly common species on our 1,700 acres. But there it was, and they saw it before I did.

Sometimes, it pays to be a child, with the fresh energy for paying attention that only children seem to have.

It's so easy to make or break a kid's interest in the natural world. When I was out with Ellie in her backyard, I casually pointed out poison ivy to two of the grandkids and told them why they needed to avoid it. Ellie told me solemnly, "Then maybe it's good not to touch any plants." I was so glad she said this, as it gave me a chance to realize I had gone overboard in my caution. Today, at age nine, she has become an enthusiastic naturalist.

But how often do our cues and cautions land on the spongelike minds of young children, turning off their curiosity and desire to explore the natural world? Sometimes, I realize what a heavy—although delightful—responsibility we have in choosing our words carefully.

Dragonfly chaser Jim Lemon remembers finding a dead Swamp Darner (*Epiaeschna heros*) in "great condition" when he was four or five years old on a family farm. "Someone in my family pronounced it a 'witch doctor.' I was advised to stay away from it." Lemon told me he snuck it home anyway, where it remained a prized possession until his mom discovered it. It sparked an interest in 4-H insect projects and other collections, leading to an MS in entomology. Lemon is now retired and chases Odes in Ohio. A lifelong interest, sparked by a dragonfly.

I've seen parents and other adults express revulsion and disgust at the idea of their children handling insects and exhibit fear if a flying bee is in the vicinity (and yes, if you have an allergic reaction to bees or your child does, this is indeed frightening). So . . . how do we balance our desire to protect kids with a need to introduce them to the wonders of the insect world around them?

There are moments when you realize you've succeeded. One of my favorites came when my daughter-in-law Gillian told me they took my grandson Jack to a spring training baseball game in Florida. Jack spent most innings pointing out the various dragonflies he saw flying around in the stadium. Although I do love baseball, I was thrilled. *Atta boy!*

Whenever I'm discouraged about my efforts in sharing the natural world with my littles, I read Rachel Carson, the brilliant marine biologist who set the world on fire with her book *Silent Spring* in 1962. It called attention to how pesticide use was wreaking havoc in the natural world. Although she had no children, when her niece died unexpectedly, Carson adopted her young grandnephew, Roger. In Carson's 1965 book *The Sense of Wonder*, published posthumously, she wrote:

> A child's world is fresh and new and beautiful, full
> of wonder and excitement. It is our misfortune
> that for most of us that clear-eyed vision, that true
> instinct for what is beautiful and awe-inspiring, is
> dimmed and even lost before we reach adulthood. If
> I had influence with the good fairy who is supposed
> to preside over the christening of all children, I
> should ask that her gift to each child in the world be
> a sense of wonder so indestructible that it would last
> throughout life, as an unfailing antidote against the
> boredom and disenchantments of later years . . . the
> alienation from the sources of our strength.

A sense of wonder. What better gift could we offer future generations? "The art of seeing has to be learned," writes Marguerite Duras. Once dragonflies and damselflies have been pointed out to a child, that person will have difficulty "unseeing" them.

It's interesting to contrast the difference in viewing insects and the natural world between a three-year-old and a six-year-old. Tony went out for a walk with me when he was three. On our walk, we passed catmint, a plant that was blooming and swarming with perhaps fifty bees. Delighted, he crouched down so he was at eye level. We talked about bees and how they collected pollen and nectar and then made honey. He was enchanted.

At six years old, he wasn't so sure about bees. An encounter with hornets had left him with several very painful stings. He'd also picked up an aversion to the occasional spider that spun a web in the corners of our house. Although he still loves dragonflies, some insects—once so fascinating—have taken on a more fearsome and negative aspect. That fresh wonder has been mitigated by negative experiences with particular insects and cues from adults who fear spiders.

When I worked at a national park as a naturalist interpreter giving bat programs, I soon realized it was important not to minimize visitors' fears. Rather, I sought to engage them with something positive about the creature they feared. One way to offer a different perspective on insects and other arthropods is to connect and engage children with their unusual habits. "Did you know spiders can taste things with their feet?" Another winner, "Let's do the dance bees use when they find flowers and want to tell the hive." Have them look for dragonflies and damselflies making "hearts" and explain their natural history.

Other dragonfly chasers and educators know the power of cool factoids. Curt Oien, an avid dragonfly chaser in Minnesota and Wisconsin, likes to share that "baby dragonflies are butt breathers and that they swim with jet propulsion." It gets their attention, he told me, as do dragonfly poop facts. *Poop facts? Tell me more!*

Ode expert Kurt Mead believes the best way to get kids involved is to give them a net next to a pond and let them loose. "The excitement of the 'hunt' and the feeling of success in catching even common dragonflies is so inspiring to watch," he said. "Most kids who are squeamish about bugs can be coaxed into holding a live dragonfly after a little supportive coaching. This small effort can go a long way toward breaking down barriers for kids who have learned to be afraid of bugs."

For other children, it begins with finding the dragonfly and damselfly nymphs. Karen Remkus, who monitors Odonates at the Morton Arboretum just outside Chicago, has worked as a naturalist guide with school groups. One of the most special moments, she said, was doing macroinvertebrate water "dips." Kids put the dipped water into trays and then looked closely at what they discovered. To the delight of the children and adults in the group, "We almost always found dragonfly and damselfly nymphs." These moments she spent with children looking at nymphs, she said, were the beginnings of her interest in dragonflies and later prompted her to become a dragonfly monitor.

On the bookshelf I keep for my grandchildren are half a dozen books on dragonflies. They are all child-friendly and written with the under-ten-years-old set in mind. But the books the kids like the best are "Nonna's dragonfly books," the field guides and oversized books with the big photos in them. I try to keep them available and not get too concerned about sticky fingers or dog-eared pages. I remind myself—it's an investment for the future.

As part of my work at the Morton Arboretum, I conducted a workshop on dragonflies for forty Children's Garden teen volunteers. We discussed the cultural history of dragonflies; some of the different species they were likely to see. And then, I talked about dragonfly mating habits. You could see them come alive! After an hour of coaching, they went to Meadow Lake nearby to look for dragonflies and damselflies on their own and attempt to ID them. Some teens goofed around; others looked bored. But several rushed up to me with their phones, showing me images of dragons and damsels and asking questions. You could see the light go on. They were hooked.

Some teenagers may be naturally passionate about insects, and you can encourage this interest. Unlike the younger set, most teens won't be won over by a pool net or a *Beginner's Guide to Dragonflies*. However, they might like a dragonfly app for their cell phones. A thirteen-year-old bug enthusiast I know taught me to use iNaturalist, an excellent phone ID app and online social network of people sharing biodiversity information to help each other learn about nature—from fungi to dragonflies to plants. I've been grateful to this teen ever since.

Once you have a snapshot of the dragonfly you wish to ID, your phone sends your location information to iNaturalist, which offers location-based ID suggestions. The ID you select is then vetted and approved by "experts" using the app. This crowdsourcing of species ID works better than you might think. I've found it to be the best of the insect apps.

I also have the Dragonfly ID app on my iPhone. It's based on the Odonata Central database, and also features crowdsourced text. It has a nice life list feature as well. If there are teens in your life who are glued to their phones, you might find that this app is also a good gateway for them to become interested in dragonflies and a way to give them more approved screen time.

Most children—and even teens, no matter how much they seem to want to avoid you—are looking for your time and your attention. How you choose to spend that time with them tells them a lot about what you value. What you get excited about shows them what is important. The future of the natural world—and its insects and dragonflies—depends on getting our children, our grandchildren, and the neighbors' kids out experiencing it and learning to love and care for it.

In her wonderful book *An Obsession with Butterflies*, Sharman Apt Russell quotes Annie Dillard, "We teach our children one thing, as we were taught, to wake up." Russell goes on to say, "Butterflies wake us up." I'd echo her thoughts here but substitute the word *dragonflies*. They help us stay alive to wonder. They remind us of the complexity and diversity of the natural world.

Rachel Carson asked, "What is the value of preserving and strengthening this sense of awe and wonder, this recognition of something beyond the boundaries of human existence?" She later answers her own question: "Those who contemplate the beauty of the earth find reserves of strength that will endure as long as life lasts."

In a world where tremendous interior strength will be needed to meet the challenges of war, violence, pandemics, accelerating technology without ethics, and political corruption, I want my grandchildren to have these reserves to draw on. Appreciation. A sense of wonder. When children develop these attributes, we may hope for a better future.

Dragonflies and insects—and learning to pay attention to the natural world—can become an important part of their lives. But it won't happen in a vacuum.

It's up to us to share the love. Pass it on.

15

Guess Who's Coming to Dinner?

The Dragonfly-Friendly Garden

Anthropocentric as [the gardener] may be, he recognizes that he is dependent for his health and survival on many other forms of life, so he is careful to take their interests into account in whatever he does. . . . The gardener cultivates wildness, but he does so carefully and respectfully, in full recognition of its mystery.

—MICHAEL POLLAN

By now I hope you're convinced: dragonflies are terrific. Now it's time to put out the welcome mat and invite them into your personal spaces. There are simple ways to make any outdoor area more inviting for these amazing creatures. Let's talk about a few.

When I moved to the Chicago region in 1998, I was dismayed by the size of my tiny backyard. Where would I put a garden? In my world, to put down roots in a place means putting in plants. Most of the yard was subdivision clay, and the only plants and trees were a few roses and some skyrocketing arborvitae, favorites of landscapers in the late 1960s when our home was built.

As a passionate gardener, I love to see the colors and shapes of a new season gradually come into focus. So immediately, I set about making my yard my own. Today, two decades later, it feels like home. From mud season in March with its bevy of marsh marigolds, through cheerful daffodils in April, then on through pink shooting star, pale purple coneflowers, and waving big bluestem in my prairie patch, the growing season is magical. And, it is made more so by the presence of the flying creatures who live or travel through my small backyard.

One thing I did to make my yard more dragonfly-friendly was put in a pond. When I worked at a national park, it was probably no surprise to my friends that my work space was an island in the middle of one of the Great Lakes. Being around water is as necessary to me as breathing. A small pond was a no-brainer.

But digging a pond in a subdivision in greater Chicago requires a procession of permits. CALL BEFORE YOU DIG! read cautionary signs across the region. First, you phone JULIE (which stands for Joint Utility Locating Information for Excavators), the independent catchall agency that coordinates notification of the various utility companies. After they visit, your lawn looks like a miniature golf course of fluorescent orange flags and multicolored spray-painted lines across the grass, designating your underground utilities. Every cable line, water pipe, telephone landline, and electrical line is revealed by a different color. Our small yard was crosshatched in every direction.

With relief, I found an area downslope by the patio that was untouched. This would be where I put my pond. At six by fourteen feet, it's small enough to manage easily. It's shallow, so that I don't have to worry about fencing it. And just deep enough. Although the surface ices over, it won't completely freeze solid in a Chicago winter. It is also somewhat sheltered from the wind, which dragonflies and damselflies both appreciate on a breezy day.

We're not talking about a fancy water space here that cost thousands of dollars. There's no plastic liner, only hard subdivision clay. A few plants I purchased—blue lobelias, cardinal flowers, and marsh marigolds—have multiplied over the years. A few other plants came in on their own: New England aster, goldenrod, duckweed, cattails. This sort of inexpensive pond has its own particular seasonal cycles. For more than twenty years, I've watched my pond overflow in a storm, then become shallow after a string of sunny days. Occasionally it will go completely dry. When that happens, I wield the garden hose and fill it, or if I'm feeling particularly virtuous, I run a line from the rain barrel and empty it into the pond.

After cancer surgery, I was ordered to rest. Walks could be no longer than ten minutes. It was August, and the garden was in full bloom. Usually when I sit on my back porch, the urge to work in the garden eventually gets me out of my chair and on my knees, pulling out invasives. But, the doctor emphasized, "No weeding!" For six weeks.

At first, it was frustrating. And then . . . liberating. I had the luxury of spending time in my backyard just observing, without the tug of guilt over the buckthorn sprouting in the garden or the garlic chives taking over the flower patch. An unexpected respite from *doing*. A reminder of what it means to be truly present. Just to *be*. As I watched the pond and the prairie patch and enjoyed seeing what new dragonfly or damselfly would show up that day, my body was able to go about the business of slow and steady healing. I wasn't focused on counting them for my monitoring work. The dragonflies were just . . . *present*. And I was present to them.

The diversity of mammal, bird, and insect life that have encountered the pond over the years is surprisingly high for a suburban puddle. Coyotes, raccoons, bunnies, skunks, deer, and foxes have all stopped by for a drink. Sometimes I surprise thirsty creatures in the early morning hours when I step outside to see the sunrise. Birds appreciate a sip or two or a bath in the shallows during the day. Two mallards, a male and a female nicknamed "Fred and Freda," drop in to paddle in circles for a bit in the spring, testing the pond for a possible nesting site (nope, too much patio activity). Red admiral butterflies check out the marsh marigolds, which necklace the pond in early April. Chorus frogs serenade us in the spring. An American toad trills on hot summer nights.

And of course there are the dragonflies. For the first time this season I have damselflies. The Eastern Forktail showed up, and—oddly enough—a Marsh Bluet, a species I've not seen anywhere else in my rambles. Ditto for the Great Spreadwing (*Archilestes grandis*), which I spotted one morning hanging out on a cattail. The Marsh Bluet, Great Spreadwing, and the Red Saddlebags I mentioned in an earlier chapter are Odes I've not seen anywhere in my more than a decade of dragonfly monitoring at natural areas. Only in my suburban backyard. Magic happens in places you least expect.

Green Darners and Black Saddlebags swing over the pond in the late summer, picking off a few wayward mosquitoes. Jeff and I pour cold drinks and sit on the back porch as the sun slants low in the sky, entertained by the antics of the Twelve-spotted Skimmer, catching the last rays of light on a lavender blazing star flowering spike, or a Widow Skimmer looping through the joe-pye weed blooms before dusk. My head clears. The tensions of the day dissolve.

In *Second Nature: A Gardener's Education*, Michael Pollan writes, "A garden should make you feel you've entered privileged space—a place not just set apart but *reverberant*—and it seems to me that, to achieve this, the gardener must put some kind of twist on the existing landscape, turn its prose into something nearer poetry." I doubt Pollan had dragonflies in mind when he penned that sentence, but I'd argue that dragonflies help turn garden prose into poetry. Their strong wings, glinting in the air. The flash of blues and reds in the sunshine. The sheer energy they bring to the space.

Dragonflies don't drink in the way that we think of downing a glass of water. I learned from Kurt Mead's excellent book *Dragonflies of the North Woods* that dragonflies absorb water through their exoskeletons and "drink" through a series of splashdowns. They also absorb dew. So think of your pond and its surrounding foliage as a drinking fountain for Odes.

When they aren't patrolling through the garden, I find dragonflies clinging to pond foliage or basking on a rock or leaf. The pond is a magnet. An invitation: "Come on in."

Your water feature might have moving water or still water. Each will attract particular species of dragonflies and damselflies. If you've got a bit of cash, you might have a stream put in. My friend Sherry has a water feature with a stream and waterfalls that empty into a small pond.

Some species like this "running water" kind of habitat. If

you use a pond liner, place some mud, gravel, or silt on top of the plastic. It will give your nymphs a place to scramble around.

Don't have room for a pond? Even half a wooden barrel with water and some plants will do. Either way, in *The Dragonfly-Friendly Gardener*, Ruary MacKenzie Dodds recommends siting your water feature in full sun. However, if you intentionally add fish and frogs, you'll increase the hazards for your flying friends. My pond has had pet store goldfish from time to time. The frogs come in on their own, probably as egg jelly masses attached to visiting mallards. I consider it all part of the circle of life and sit back and enjoy the show.

What about mosquitoes? When you have standing water in your yard, they can become an issue. I mitigate it in three ways.

First, a solar bubbler keeps water moving but doesn't require electricity or power cords and expensive equipment. It's not foolproof, but it costs less than fifty dollars, so if I do have a problem with it (say, the pond dries unexpectedly while I'm away and the battery burns out), it is replaceable without too much hand-wringing on my part.

My second strategy: attracting dragonflies to my yard, which will help cut down on the mosquito population.

Third, if I have a particularly severe mosquito season, I use *Bacillus thuringiensis israelensis* (Bti), which is a specific-target biological pesticide and appears to be nontoxic to dragonfly nymphs (from current research at this writing), fish, pets, and wildlife. Doughnut-shaped Bti "mosquito dunks" are dropped into standing water and biodegrade over time. They release a *larvicide*, a substance that kills the mosquito larvae before they become biting adults. Mosquito dunks are my last resort. I opt not to use them, proven as they are, unless they make the difference between spending time in the backyard and not.

What about plants? There isn't a lot of research on plant associations with dragonflies, unlike monarch butterflies, with their iconic affinity for milkweed. Dragonflies use plants as perching spots or as grocery stores full of insects that they can pluck and enjoy. A diversity of foliage size in your pond plants is likely to increase the diversity of species you'll see in your dragonfly and damselfly visitors. Plan for some tall plants in and around the water to give the dragonflies and damselflies a place to perch.

Some Spreadwing damselflies cut slits in cattails and other pond foliage to oviposit—lay eggs—above the water surface. This method differs from that of most dragonflies, which oviposit by tapping or laying eggs directly into the water or into floating vegetation mats. Just think! Your pond becomes a nursery for dragonflies and damselflies, ensuring that you'll have future generations of Odes to enhance your yard.

Dragonflies need a place to rest and sun themselves, so offer them a few rocks large enough for basking. Lily pads are a good resting perch for a dragonfly if your pond is large enough. Dragonflies will use your pond as a place to recharge their solar batteries, and you'll be able to get an up-close look at them as they do.

Pesticides, if you must use them at all, need to be kept away from the pond. I occasionally remove buckthorn plants that threaten to take over the garden, and when I do, I dab the pesticide directly on the offending plant and remove it. If you spray herbicide, you take a chance of hitting nontarget plants. Keep all chemicals out of the water. If you're using lawn fertilizer or weed killers broadcast by a lawn care company, save yourself some money and discontinue the service. You'll have a healthier yard, with a greater diversity of insects, for yourself, your children and grandchildren. The poison doesn't do anyone any good. Dandelions? They make excellent fritters when battered and fried.

As you're planning your dragonfly garden, keep in mind that dragonflies are not pollinators. If they do pollinate something, it's likely accidental. In fact, dragonflies eat some of our pollinators. Wait—don't close this book! It's not bad, I promise. Remember, beautiful garden birds such as hummingbirds, which are pollinators, also eat other pollinators. A rich and diverse world requires that *eat-and-be-eaten* arena. Frogs eat dragonflies. Dragonflies eat mosquitoes. It's all part of the natural community.

All pollinators are not created equal. Did you know that mosquitoes and wasps are pollinators? And while you might have a MONARCH WAY STATION sign in your back-yard as I do, you likely won't have a sign for beetles, which are the stronger pollinators. Most of us aren't gardening to attract those insects, nor are we putting out the welcome mat for wasps and mosquitoes. And even though dragon-flies are not pollinators, they may benefit your garden by eating insect pests, including those same pollinating mos-quitoes. Our natural world is a complicated place. Give yourself a ringside seat by making your yard a dragonfly-friendly spot.

One hot summer evening, I sit in my lawn chair and watch a Green Darner scout insects as it flies across the lawn. Occasionally, the Darner snatches an unwary mos-quito. *Thank you*, I murmur. Unlike the vegetable garden or bird feeders, which all require plenty of maintenance, the dragonflies demand little of me except some water and my occasional attention. And even if I forget to fill the pond or notice them, they'll continue with their lives, unconcerned.

There's something restful about that.

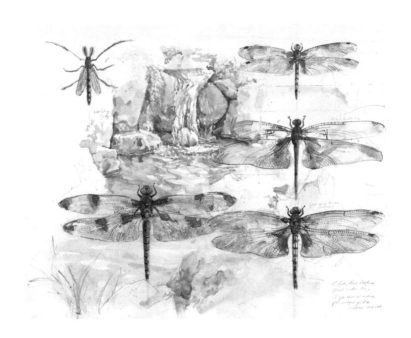

16

Ode Addiction

A Cautionary Tale

Every one is inclined not only to promote his own study, but to exclude all others from regard, and having heated his imagination with some favourite pursuit, wonders that the rest of mankind are not seized with the same passion.

—SAMUEL JOHNSON

*B*eetles and wasps don't make it onto the front of too many greeting cards. But, during my cancer recovery, my mailbox overflowed with cards full of get-well wishes and dragonfly paintings, dragonfly photos, and—most beautiful of all—a card with a pop-up, laser-cut dragonfly. Dragonflies have the advantage of a certain cultural aesthetic.

It's this aesthetic and passion for Odes that leads us dragonfly chasers down a dark and narrow road. You find yourself steering conversations to topics about insects. Pulling over by streams running through neighborhood subdivisions "just to take a look." Planning your overseas trips around the best places to see dragonflies. Diverting your business lunch date to a dragonfly destination or luring your spouse to a particular location ("I heard there is a great restaurant in this town along the river—we should check it out!").

Chasing dragonflies, like buying good books or eating gourmet food, can be curiously addicting. Before you know it, you are putting down hard cash for all things dragonfly, while talking about your "free" hobby. And hey—is that a dragonfly shower curtain in your bathroom?

I shared some descriptions of my dragonfly hoard—I mean, collection—earlier. I'm not alone. We live in a consumer-driven culture, where anything we have a passion for is packaged and stylized for mass-market consumption. Mugs. Earrings. Bookmarks. Handbags. Wallets. Pottery. Clothing items. Sometimes I feel happy just surrounded by it all. I don't even need to go outside . . . do I?

In William Cronon's 1995 book *Uncommon Ground: Rethinking the Human Place in Nature*, he includes an essay from Jennifer Price called "Looking for Nature at the Mall." In it, she examines the Nature Company retail stores, and their (now former) place in American culture. In case you're not familiar with the Nature Company, it was a chain of retail stores—founded in 1972 and largely located in malls, airports, and shopping plazas—that specialized in the idea of nature as presented through audio recordings, toys, art, music, and other retail items. Each store had a unique "water feature." By 2001, all 114 stores were either closed or transformed into Discovery Channel stores.

I remember one of these stores in the Chicago suburbs. I shopped there with enthusiasm! The store's carefully calibrated surroundings, the music, the items available for purchase all made me feel good. It was a lot like being outside—without any of the attendant hazards of actually *being* outside. No biting flies. No turned ankle from stepping into a crevice on a trail. No hot sun causing my shirt to stick to my back and sweat to run between my shoulder blades. The air-conditioned store spoke to something I aspired to and, it's not too strong to say, something I yearned for.

As a suburban mom at that time, I didn't know quite how to satisfy that yearning. I gardened heavily. Took natural history classes. I wanted that feeling of being a part of the natural world. But I wasn't always sure how to *be* in the natural world.

"We've filled our homes and offices with images of nature from everywhere," Price observes. What we hunger for, she says, is a sense of place: "In this fast-paced, ever-changing world [and remember, she's writing in the 1990s, before cell phones and before the internet exploded into common use] we count on nature not only to stay constant in meaning but to stay put. A poster of the Antarctic or the Amazon rainforest inspires, among other things, a spirit of place." She adds, "Far from the ocean, the plastic whale reduces easily to a motif, a feeling, an association—like 'freedom' or 'beauty,' or perhaps 'solitude.'" It's easier that way, than seeing a whale in real time, on a lurching boat, perhaps in the pouring rain. "Why are we looking for nature with our credit cards?" she then asks. She quotes a *New York Times* writer, "Is it possible that people in our culture have become so estranged from nature that their only avenue to it is consumerism?"

It resonates. I've found myself substituting facsimiles of nature for nature itself. At home, it begins subtly. Usually it starts with the bathroom, a favorite spot for nature motifs. A dragonfly-themed shower curtain here, a dragonfly-imprinted soap dispenser there. Then, "The photo of the Halloween Pennant dragonfly you took would look great framed—let's hang it up in the hall!" Next shopping trip, "Wow, check out those terrific embroidered dragonfly hand towels. And they're on sale." A book of dragonfly photography appears on the coffee table. You head for work and reach for your dragonfly-emblazoned thermal carafe. A dragonfly sticker appears on your car's back bumper. Then another. And one more.

Suddenly, it hits you. Your home is a repository for all things dragonfly. It happens so gradually, you don't even realize it until one day when a visiting friend, bemused, asks you, "So what's with the bugs?" Or you're sitting in the garden, reading, and you hear the wind chimes—the dragonfly-themed wind chimes. You put your dragonfly bookmark into your new book on Odonates and take a sip from your dragonfly mug as you soak up the ambience.

To add insult to injury, another visiting friend, this one a scientist, comments on the improper number of segments on the shower curtain dragonfly's abdomen, or the wrong positioning of the damselfly's wings on the soap dispenser. *Come on! It's a soap dispenser, not an anatomical rendering. Haven't you heard of "stylized"?*

I mentioned books. Oh, those dragonfly books. They "spark joy," dozens of them, and there are more on order. Natural histories. Cultural histories. Photography books. A volume of haikus—all of them about dragonflies. Children's books. ("For the grandkids," I said. Sure they are.) Dragonfly field guides for the countries we travel to: Ireland, England, Italy. Books on dragonflies in particular regions or states, such as the "North Woods," Wisconsin, or Ohio.

It could become a habit–to be indoors, paging through my dragonfly books, surrounded by my dragonfly housewares, and never experiencing the knowledge or connection I get when learning an ID standing ankle deep in the mud or wading through the tallgrass in hot pursuit of a just-missed-it Darner of some sort. Making that connection of the heart.

I like the "idea" of dragonflies. I enjoy having their images around me. Writing about them gives me deep pleasure. Showing images and giving talks about them to civic groups and conservation organizations is deeply satisfying. The encounters with dragonflies and damselflies in real time, however, bear small resemblance to the encounters with my shower curtain dragonflies, or even writing or talking about them. While all are satisfying, nothing replaces that moment in the field. *Nothing.*

Field time is more difficult. Sunburns don't happen in your living room reading a dragonfly guide. Nor do you have to swat biting insects while using a soap dispenser or donning a pair of dangly dragonfly earrings. But it's those moments outdoors when you relax. Stress fades. Epiphanies happen.

As I spend time with my six grandkids, one of my goals is to invite them to pay attention to the natural world. Fortunately, my two adult children, who in junior high bitterly protested nature hikes, have become wonderful parents who love and cherish the outdoors. So we're in this together. Sure, I buy the little ones plastic frogs or a stuffed turtle. We fly a butterfly-shaped kite together, or enjoy a trip to the Field Museum. All good things. But the challenge for their parents and me is to help them to move away from the facsimiles and engage in real time with a world they see shown on *Wild Kratts* or *Nature Cat*.

I'm grateful when parents do everything they can to get the kids outdoors. It requires an investment of time. And a whole lot of dirty laundry and muddy shoes. All relationships take time and energy. The result? A heart connection as well as a mind connection.

When children have this connection, we can rest assured. The future is in good hands.

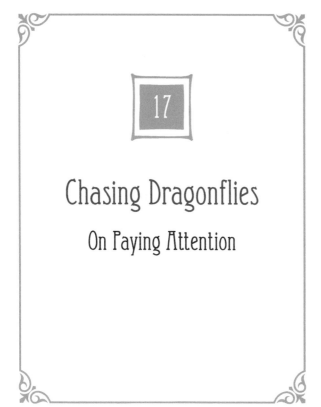

17

Chasing Dragonflies

On Paying Attention

Pay attention. Be astonished. Tell about it.

—MARY OLIVER

It's a beautiful day to be outside, blowing bubbles. The July sun warms the wooden bridge over Willoway Brook, taking the chill off of an early summer morning. The humidity is a whisper in the background; soon, it will smother everything like a damp blanket. I dip the plastic wand into the bottle of soap bubbles and gently coax translucent orbs from the circle. They spill into the air, catch the breeze, and float across the prairie. Each one a miniature reflection of the world.

I have a lot of soap bubble bottles around. One of the fallouts of having six grandchildren is you have an excuse to keep fun things in the house that might otherwise look odd for someone my age: sidewalk chalk, Lego bricks, a library of children's literature (which I read and reread under the pretext of keeping them for "the kids"). And the bubbles.

A light wind has sprung up, tousling the black walnut tree leaves by the bridge and sending a ripple of waves through the purple prairie clover and the young prairie grasses. It's quiet here; most people stick around the visitor center at the Morton Arboretum. A tallgrass prairie is a good bet for some alone time. I coax a few more bubbles from the plastic wand. They catch the air currents and bounce over the tallgrass, flash sunlight: blues, pinks, and golds.

After I amuse myself with this for a few minutes, I notice that the dragonflies are agitated. A male dive-bombs the globes. Does he see these bubbles as some sort of alien species of dragonfly, invading his streamside property?

I put down the bubble wand and lie on my back, watching the clouds sprawl across the sky. The tallgrass whispers, smelling of chlorophyll and wet earth. Below, the stream ripples on its way, the sloshy sounds of water jostling rocks. Overhead, bluebirds flit from walnut tree to bur oak; blue on blue sky. The anxieties of the day retreat.

In her poem "The Summer Day," Mary Oliver asked, "Tell me, what is it you plan to do with your one wild and precious life?" That question has haunted me since the day I first read the poem, years ago. I don't know exactly what I plan to do. But this much I know: I want to be here, present in the moment. I want to pay attention.

The more I pay attention to the natural world around me, the more I learn about myself and the world I'm part of. Paying attention to the natural world might include planting herbs in my backyard garden, backpacking in the wilderness, monitoring dragonflies, walking at a nearby arboretum, or sitting on the bridge here at this tallgrass prairie. Nature, wherever I find it, lets my soul breathe. It's where I let go of some of my doing and just *be*.

It's not easy. Although there was a fine tradition of porch sitting and watching the world go by in generations past, today there is little virtue conferred on those who pause to enjoy a sunset, or spend an idle hour watching dragonflies, or walk for anything other than exercise.

And, of course, there is less affirmation for sitting on a July morning, blowing bubbles.

Paying attention began for me with journaling what I saw, a veritable listing of "who what where when why." I began consciously using my five senses. Inhaling. Feeling the difference between one leaf and another. Trying to understand why the damselflies were locked in a heart-shaped wheel, wings fluttering, in the brook that ran through the prairie. My old journals are filled with descriptions: the smell of decaying leaves, the prickly softness of the saffron and black woolly bear caterpillar, the rasp of a grasshopper in the tallgrass, the cool taste of mountain mint leaves.

These sensory observations draw me into my interior landscape. Sometimes they prompt memories. When I discover tiny ruby strawberries half hidden in the spring grass, then pop them into my mouth, my grandmother's face comes unbidden to my mind. Grandma's kitchen was filled with strawberry motifs on canisters, plates, teacups, and dish towels. The floor was laid in alternating green and red linoleum tiles, and the walls were papered with a strawberry pattern. Grandma had a complete wardrobe–right down to earrings and a pair of shoes–based on the berry. I hunt the strawberries in the grasses, then taste them. Tangy and sweet. I remember her.

A mourning dove calls as I pull weeds in my backyard, and I think of how I learned its name. While other girls received Barbie dolls as gifts, my maternal grandmother–a science teacher–gave me rock tumblers, a microscope, and fishing trips on the lake. She taught me the names of the wildflowers, trees, and birds like this one. I listen to the dove and it all rushes back. The taste of the sandwiches Grandma packed for our lunch. The bluegill we caught and fried later in her kitchen.

Paying attention makes room for these memories. Not all of my memories of my grandmothers are pleasant. But memory has a way of opening doors of understanding about where I came from and who I have become. They remind me of how I was shaped; they help me reflect on where I'm going with my life. Memories offer context. They help me make course corrections on my journey.

If you don't want to remember, stay busy. To be still without doing–to open up and pay attention in nature–is a risk. The memories will come. In all memory lies the prospect of pain.

A friend heard of a three-week solo backpack trip I was about to take to Isle Royale, a wilderness island in Lake Superior, where wolves and moose freely roam. "It's not the wolves I'd be afraid of," she mused. "It's being alone with my thoughts so long." Fear of memories trumps wolves, hands down.

So instead of being in nature and opening ourselves up to what comes, we find substitutes that aren't so frightening. We watch *Animal Planet*. Put on dragonfly earrings. Page through a book about a backpacker's epic hike. We listen to bird songs online, then go to a meeting about global warming. These are all good things—but they are no substitute for the natural world. The *real* things.

What are the real things? While browsing through the Sunday newspaper, I found an article about an exhibit of John James Audubon's classic bird portraits. I read on, intrigued. Then I came to this line: "Many of us will never see a great horned owl . . . but these representations more than suffice."

Oh, really?

The great horned owl is ubiquitous in my area. I've spent many happy evenings slipping through a grove of red pines on the city's outskirts, listening to its call and enjoying the owl's crisp silhouette on a treetop at dusk. When I walk in the springy duff under the pines, I find owl pellets. Picking the pellets apart, I discover feathers and bones and hair, all evidence of a late-night meal. Sometimes I spot owls in the daylight, mobbed by a band of crows.

These experiences of looking, paying attention, and discovery on my walks are things that no artwork could ever replace. Not even Audubon's spectacular portraits of birds.

"New actions inevitably lead to new experiences, and this novelty kicks our senses out of the slump induced by routine," writes Tristan Gooley in *How to Read Nature: Awaken Your Senses to the Outdoors You've Never Noticed*. "This in turn raises our levels of awareness. One of the peculiar consequences of starting to notice new things is that we cannot help but notice how little we have been noticing. . . ."

At the arboretum where I walk, I regularly dodge cars that whiz by, windows rolled up tight and doors locked. Walkers pass me, pumping weights, glued to their cell phones, or listening to music with their earbuds firmly in place. Sometimes they juggle all three. Few people come to the arboretum to hike in the winter; however, thousands flock to it for programs or special events indoors in the colder months. Having an unplanned agenda is anathema to many. We want to know what to do. And we want to "do it" in a place that is comfortable and doesn't demand much from us.

Insects gave me a window into this disconnect. Several years ago, the Chicago metro area experienced the cyclical advent of the seventeen-year cicadas. Casual visitors to the arboretum were clueless about how to handle the sudden onslaught of insects, a veritable plague of biblical proportions. All of the tricks people use to isolate themselves from the more inconvenient aspects of nature were useless.

As I'd be out looking for dragonflies, cars would pull over. Questions, shouted over the shrill ear-splitting drone of the cicadas, came thick and fast: "Where are the cicada-free areas?" Others asked, "Where are the cicada viewing stations?"

The cicadas, of course, didn't fit into any neatly conceived program or box. They weren't corralled into a handy viewing station where you could buy a ticket and see them behind a glass window. Cicadas flew everywhere; noisy, inescapable, and . . . beautiful, if you paid close enough attention. In their own, inimitable, cicada-like way.

Another early morning, I wandered inside the Morton Arboretum's visitor center for a cup of coffee. At the front desk, a mother with two young children was venting her wrath at a volunteer because the gated Children's Garden play area didn't open for another thirty minutes. As I walked past, she shouted, "Well, what am I supposed to do with these kids for the next half hour?" Really? The arboretum is a 1,700-acre area with a diversity of trees, wildlife, ponds, and prairie. Miles of hiking trails. You could spend an hour sitting in the grass, watching the insects and birds and squirrels. But for this woman, free play or strolling aimlessly with her children wasn't an option. She had an agenda. She couldn't think outside of it.

Programs. Structure. Agendas. Comfort. Our desire for the orderly and the predictable often crowds out our ability to be spontaneous and to immerse ourselves in the moment. To pay attention. To remember. To embrace the mystery of the short time we have in this world.

Waiting quietly or "doing nothing"—even blowing bubbles and watching dragonflies on a lazy afternoon—may counter our desire to be useful or to accomplish something.

And yet.

When you pay attention, you prop the door of your soul ajar to welcome the unexpected and the uncontrolled. You do nothing. You stay open to receive. But receive what? You aren't sure. You can't access your interior landscape by banging down a closed door; you can't pencil it in as an appointment on your calendar. Paying attention is a habit-forming mind-set that comes with repetition and with intention. You give yourself permission to "do nothing." You create quiet spaces. You open a door.

On my solo backpacking trips, my greatest fear is getting lost. In the worst places on the trails—or lack of trails—I look for piles of rocks called *cairns*, built and left behind by other hikers, that point me in the right direction. When you begin to learn to pay attention, there are many cairns—generous gifts that are given without expectation of repayment—that tell you that you're on the correct path.

We know we are paying attention when we slow down. We breathe more deeply. Our minds stop whirling. The problems that seemed insurmountable are seen as relatively insignificant. We then return to the demands of our day, refreshed and better able to cope. We're more patient. We're not so apt to snap at our loved ones.

When I'm out chasing dragonflies, I slow down. I look closely. I pay attention.

And paying attention attunes us to wonder. We look around us and are astonished at the complexity of a leaf, the gymnastics of a squirrel, the grandeur of an approaching thunderstorm. We notice a delicate, yet strong spider-web glistening with dewdrops. We discover a maple tree in spring, dripping icicles of sweet sap, and we taste and take the very essence of tree inside of us. Later, we stand under the stars, marveling at the Milky Way galaxy, splattered like raindrops across the dark windshield of the universe.

Each discovery is another "cairn" that tells us we are moving in the right direction. With these observations of "ordinary" wonder comes awe, swiftly followed by gratitude. Our perspective shifts. Our glass moves from half empty to half full.

"To pay attention: This is our endless and proper work," wrote Oliver. Paying attention also reminds us of the value of spontaneity. It helps us stop pressing our own agendas. We learn to listen. We learn to be flexible. We learn to let go. We don't try so hard. It's okay to let others call some of the shots. We don't have to always be in charge or the center of attention.

When my health fell apart, I learned a lot about who I was apart from the labels that I had previously called my "identity." All I could do was wait. Be still. Watch the dragonflies from a lawn chair. Heal. And learn more about my interior life, devoid of the busyness that had distracted me from it before. Who was I, apart from all the trappings? Would it be okay if I never wrote another book or taught another class?

Paying attention reminds us of the vast diversity of the world. It pulls us out of our tight boxes and opens us up to new ways of seeing. Nature gives us a sense of something bigger than ourselves, and a reminder of how much we don't know or understand. We become more empathetic, more understanding, less judgmental. We relax into whatever place or circumstance in which we discover ourselves. We leave room for mystery.

Amazed and humbled, we crack the door of our lives open a little wider. We walk, and we watch, and we look for cairns to guide us. We aren't afraid to "do nothing" and wait. Instead, we blow bubbles and enjoy the dragonflies.

Open for whatever comes.

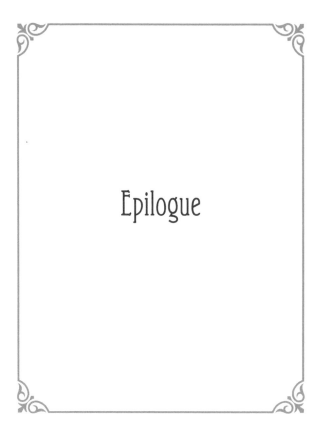

Epilogue

Let us go on, and take the adventure that shall fall to us.

—C. S. LEWIS

Spring. We burn the tallgrass prairie to rejuvenate it for a new year. The first fuzzy lavender pasque flowers bloom. Woodlands erupt in white large-flowered trillium, pinky-purple wild geranium, and bluebells in generous pools of color, seemingly reflections of the clear skies above. Squirrels race around my porch, crazy with the warm weather. The mallard ducks are back. They land in my small backyard pond, necklaced with marsh marigolds in full bloom.

Along the stream that runs through Willoway Brook there's a flush of emerald growth. Red-winged blackbirds rasp their "Ok-a-leeeeee" and eye-popping bluebirds skim the edges of the prairie. Shoots of violets, prairie alum root, and rattlesnake master show their distinctive leaves in miniature as they push up through the newly warmed soil.

As the ice disappears from the stream, I know it's only a matter of time before the first Green Darner migrants arrive from the south. Soon, Ebony Jewelwing damselfly nymphs will pull themselves out of the water and trade their old lives as creatures of the water for the life of the air. I scan the skies, wondering which species will be the first I see this season. Green Darner? Black Saddlebags? I think of Scott King's haiku:

> The first dragonfly of the year
> almost too fast
> for out-of-practice eyes

It's a curious thing, to become involved in the natural world. Your perspective changes. You're aware of the brevity of life; the vulnerability of the tiniest insect. Nearing sixty, I'm aware that each year is something to be cherished. Cancer also reminds me of this.

My family knows that one of my pet peeves is hearing the phrase "killing time." And God forbid someone tell me, "I'm bored." Time is precious. The world is a fascinating place. How could anyone be bored? Look around and there are a hundred amazing things to investigate. Open your eyes and you see clouds build and swirl, insects buzz by. Sit still long enough in the spring and you can watch a blood-root bloom before your eyes. Examine a crack in a sidewalk and marvel at the busy work of the natural world going on under and around the concrete.

Pay attention. Be astonished. Tell about it. For years, I've tried to live out this simple philosophy espoused by the late poet Mary Oliver. Since my cancer diagnosis, I feel a renewed sense of urgency to share the natural world. Paying attention to dragonflies and damselflies is a way of practicing focus. It's not for the impatient or the person who eschews discomfort. Chasing dragonflies doesn't cost a penny, but it does require an investment of time. It requires patience. A willingness to wait and then, to see what comes. An openness to mystery. A resistance to "killing time." Instead, a desire to invest your time wisely.

Seeing dragonflies is a lifelong journey. They will show up when you least expect it. You'll be aware of them in parking lots, at baseball games, or while eating at an outdoor restaurant. In the winter, you'll walk by frozen creeks and wonder what will emerge in the spring.

For most of my life, I lived unaware of dragonflies. Oh sure, we coexisted. But I didn't truly *see* them as individuals or as distinct species. There was so much waiting to be discovered, right under my nose. *Dragonflies. Damselflies.*

Because of dragonflies, I'm more aware of what happens when I pollute the water with my plastics or the air with my car exhaust. I'm cognizant of the preciousness of our water and the fish, mammals, and insects that depend on it for life. I'm aware that what hurts them will eventually hurt me—my family, my friends, my grandchildren. I'm aware that when this river is channeled, or that wetland is destroyed, the consequences will reverberate long past my lifetime.

Because of dragonflies, I care more.

Following the rhythm of the dragonflies through the seasons has been an exercise in paying attention. In learning to slow down. Be quiet. Listen. Look. My old ways of seeing the world have changed.

How will dragonflies be a part of your life? I hope this book will be a spark that kindles a lifelong interest for you in Odes, if you didn't have an interest before you came to these pages. Perhaps your interest begins and ends here—and that's okay. I hope you'll have a heightened awareness of dragonflies and notice them whenever you are sitting at a traffic light or walking alongside a lake or pond. Being attentive to what is around us in the world is a gift in itself. The dragonflies and damselflies will reward your attention with their beauty, fascinating aerial dynamics, and endless diversity.

You may decide to become a citizen scientist and learn to monitor dragonflies, collecting data that will help researchers chart the course of their migration and land stewards make decisions about the care and management of the natural world in the years to come. If so, welcome to the club. You'll find a wealth of information in the resources listed at the back of the book to help you as you continue your quest.

Perhaps you're a dragonfly expert, or already hooked on dragonflies, or maybe you research or monitor or teach classes about dragonflies. For you, reading this book is just another episode in your ongoing love affair with a charismatic insect. My wish for you is that you'll be encouraged as you enter data and contribute to our body of information about these fascinating fliers.

If nothing else, you'll have absorbed enough dragonfly information to share it with the children in your life. Because of you, they'll grow up in a world where they are aware of dragonflies as something special, something to be cherished and protected. It's a gift you can give to the future and a way to make a difference in a world where insects are vanishing at an alarming rate.

Annie Dillard writes, "How we spend our days, of course, is how we spend our lives." I've been chasing dragonflies in one form or another since 2005, and every year is a new experience. There is a lot of mystery in the world. I have a lot of questions.

Especially now. Cancer has been a wake-up call. I'm fortunate to have a good prognosis; I spend my doctor's appointments sitting in waiting rooms with many who do not. But all of our days are finite: like the dragonflies'. Good health is no guarantee of longevity.

In her poem "When Death Comes," Oliver wrote, "When it's over, I want to say: all my life I was a bride married to amazement." Chasing dragonflies encourages us to dive deep. Appreciate beauty. Practice gratitude. Marry amazement.

So much about the future is unknown. One thing I'm sure of: however I spend the time left to me—months, years, or several decades—I want to live intentionally. To be awake and aware. The dragonflies help me do just that.

I can't wait for the next season to begin.

Spring comes—
the dragonfly is back
on its path—

—KEN TENNESSEN

Favorite Dragonfly-Chasing Guides

This is not an exhaustive list for midwestern dragonfly chasers, but it does include some of my favorite resources. Enjoy!

Beginners Guide to Dragonflies and Damselflies by Blair Nikula, Jackie Sones, Donald Stokes, and Lillian Stokes. New York: Little, Brown, 2012. When they turn three, my grandchildren get this book, which they immediately decorate with stickers. A favorite bedtime read. Fits in their coat pockets and is easy to use. Adults, such as my beginning dragonfly monitors or casual enthusiast students, like it too.

Damselflies of Chicagoland by Marla Garrison. Chicago: Field Museum, 2011. https://fieldguides.fieldmuseum.org/guides/guide/388. This free digital guide is regional, specific, and invaluable. Marla is a treasure.

Damselflies of Minnesota, Wisconsin, and Michigan, 2nd ed., by Robert DuBois. North Woods Naturalist Series. Duluth, Minn.: Kollath+ Stensaas Publishing, 2019. You will love having this guide in the field to help you decipher who's who in the damselfly world.

A Dazzle of Dragonflies by Forrest Mitchell and James Lasswell. College Station: Texas A&M University Press, 2005. A treasure trove of dragonfly cultural history, beautifully illustrated. Dazzling!

Dragonflies and Damselflies: A Natural History by Dennis Paulson. Princeton, N.J.: Princeton University Press, 2019. Gorgeous, current, and informative, it belongs on every dragonfly chaser's bookshelf.

Dragonflies and Damselflies of Northeast Ohio, 2nd ed., by Larry Rosche, Judy Semroc, and Linda Gilbert. Cleveland: Cleveland Museum of Natural History, 2008. Out of print at this writing but well worth your time to look for a used copy. Try for the spiral-bound edition, if you can find it. Indispensable and offers broader use than the regional title implies.

Dragonflies and Damselflies of the East by Dennis Paulson. Princeton, N.J.: Princeton University Press, 2012. It's difficult to find such a broad area represented more beautifully than in this guide.

Dragonflies: Magnificent Creatures of Water, Air, and Land by Pieter van Dokkum. New Haven, Conn.: Yale University Press, 2015.

Dragonflies of the North Woods, 3rd ed., by Kurt Mead. North Woods Naturalist Series. Duluth, Minn.: Kollath+Stensaas Publishing, 2017. Nice guy, great field guide, wonderful updated edition. Even if you have the first or second edition, it behooves you to buy the third. In fact, this is an amazing series (North Woods)—check out the complete library. You'll be hooked, as I am.

Dragonflies Q&A Guide: Fascinating Facts about Their Life in the Wild by Ann Cooper. Mechanicsburg, Penn.: Stackpole Books, 2014. A terrific book to acquaint yourself with basic dragonfly lore; I recommend this one to my students who don't want to focus on identification but want to learn dragonfly factoids.

Dragonflies through Binoculars: A Field Guide to Dragonflies of North America by Sidney W. Dunkle. New York: Oxford University Press, 2000. The first field guide I bought about dragonflies, now out of print but still available used.

Dragonflies by Cynthia Berger. Mechanicsburg, Penn.: Stackpole Books (Wild Guide), 2004. Out of print, but still available used.

Illinois—Common Dragonflies and Damselflies of the Chicago Region by the Field Museum Volunteer Stewardship Network and Chicago Wilderness. Chicago: Field Museum, 2006. https://fieldguides.fieldmuseum.org/guides/guide/380. This free downloadable four-page guide contains full-color photos to aid in identification.

For Children

Dazzling Dragonflies: A Life Cycle Story by Linda Glaser. Minneapolis: Millbrook Press, 2008.

Dragonflies by Suzanne Slade. New York: PowerKids Press, The Rosen Publishing Group, Inc., 2007. (Under the Microscope: Backyard Bugs)

Dragonflies: Catching, Identifying, How and Where They Live by Chris Earley. New York: Buffalo, Firefly Books, 2013.

Helpful Web Links and Organizations

Dragonfly Society of the Americas
https://www.dragonflysocietyamericas.org/

Illinois Dragonfly and Damselfly Checklist
http://www.museum.state.il.us/ismdepts/zoology/odonata/

The Illinois Odonate Survey
https://illinoisodes.org

Migratory Dragonfly Partnership
http://www.migratorydragonflypartnership.org/index/welcome

Odonata Central, Dragonflies of the Americas
https://www.odonatacentral.org

Odonate Videos by Mark Donnelly
https://markdonnellyphotography.org/dflies

Worldwide Dragonfly Association
https://worlddragonfly.org

Xerces Society for Invertebrate Conservation
https://xerces.org

Blogs

Tuesdays in the Tallgrass @Wordpress, a prairie blog by the author, often contains photos and information about dragonflies and damselflies in Illinois.
https://tuesdaysinthetallgrass.wordpress.com

Species of the Month with Dragonfly Society of the Americas focuses on a different species each month and includes incredible photographs.
https://www.dragonflysocietyamericas.org/en/blog

Social Media

Facebook and Instagram: Dragonfly Society of the Americas

Facebook: The Illinois Odonate Survey

Facebook: Midwest Odonata

Facebook: Odonata of the Eastern United States

Twitter: British Dragonfly Society (@BDSdragonflies)

Twitter: Dragonfly Society of the Americas (@OdonataAmericas)

Twitter: Worldwide Dragonfly (@WorldDragonfly)

Cancer Information

For information on uterine cancer, please explore the following resources and share them with your mothers, your daughters, your sisters, your girlfriends, and the women in your life:

Mayo Clinic, "Endometrial Cancer," https://www.mayoclinic.org/diseases-conditions/endometrial-cancer/symptoms-causes/syc-20352461. Helpful information presented by the Mayo Clinic includes warning signs of endometrial cancer, treatment, prevention, and when to see a doctor.

100 Questions and Answers about Uterine Cancer, by Don S. Dizon and Linda R. Duska. Sudbury, Mass.: Jones and Bartlett Publishers, 2010.

Patients' Guide to Uterine Cancer, by Teresa P. Diaz-Montes, MD, MPH, Johns Hopkins Medicine. Sudbury, Mass.: Jones and Bartlett Publishers, 2009.

I owe a debt to the following dragonfly chasers' written works (plus a few others not mentioned here). I've absorbed their books, words, and images in so many ways, and I can never thank them enough for their "virtual" mentoring of this dragonfly chaser and naturalist through their publications. Grateful thanks especially to writers Kurt Mead, Dennis Paulson, Marla Garrison, Robert DuBois, Ann Cooper, Forrest Mitchell, and James Lasswell. I reference their books and guides specifically in the notes that follow (see the preceding "Favorite Dragonfly-Chasing Guides" section for full citations of sources that are abbreviated in the notes), and any omissions in attribution are unintentional.

Opening Epigraph

vii *"It seems to me . . ."*
David Attenborough quoted in Mark Thompson, "Sir David Attenborough Is an Inspiration–And These Quotes Prove It," iNews, last updated September 6, 2019, https://inews.co.uk/inews-lifestyle/people/15-times-david -attenborough-538895.

Prologue

xiv *"It might have been otherwise."*
Jane Kenyon, *Otherwise: New and Selected Poems* (Minneapolis: Graywolf Press, 1996), 214.

xv *"There is no reason to write a book . . ."*
Richard Manning, *Grassland: The History, Biology, Politics, and Promise of the American Prairie* (New York: Penguin Books, 1995), 1.

xvi *"All this existed before . . ."*
Sharman Apt Russell, *An Obsession with Butterflies* (New York: Basic Books, 2003), 1.

xvii *"Days are short—"*
Kobayashi Issa, "Days Are Short," in Kobayashi Issa,

Ken Tennessen, and Scott King, *Dragonfly Haiku*
(Northfield, Minn.: Red Dragonfly Press, 2016), 42.

Chapter 1

4 "*Today I saw the dragon-fly . . .*"
Alfred Lord Tennyson, "The Two Voices" (1883), in
The Poetic and Dramatic Works of Alfred Lord Tennyson
(New York: Houghton Mifflin, 1898), 30. I was first intro-
duced to this poem in *Dazzle of Dragonflies*, 29.

6 *"insect apocalypse"*
Brooke Jarvis, "The Insect Apocalypse Is Here," *New
York Times Magazine*, November 27, 2018, 45, https://www
.nytimes.com/2018/11/27/magazine/insect-apocalypse.html.

8 *the order Odonata, which simply means "toothy ones"*
"Introduction to the Odonata: Dragonflies and Damselflies," University of California Museum of Paleontology
(website), accessed October 24, 2019, https://ucmp
.berkeley.edu/arthropoda/uniramia/odonatoida.html.
Marla Garrison (email, May 23, 2019) was extremely
helpful to me with her detailed explanation: "Dragonflies
and damselflies have sharp pointed hooks. There are four
jaw parts—the labrum, mandibles, maxillae, and labium.
The mandibles have sclerotized, (or hardened) 'teeth'
referred to as their molars and incisors."

9 *At one time, earthworms . . .*
Edwin Way Teale, *The Strange Lives of Familiar Insects*,
(New York: Dodd, Meade, 1962), 18–19.

9 *oldest insects in the world . . . Devonian*
"Oldest Insect," Guinness World Records, accessed
October 24, 2019, http://www.guinnessworldrecords.com/
world-records/oldest-insect. Marla Garrison was a great
help in this section on understanding the early forms of
Odonata (email, June 3, 2019).

9 *ancient relatives of the dragonflies, the Protodonata*
S. R. Jongerius and D. Lentink, "Structural Analysis
of a Dragonfly Wing," *Experimental Mechanics* 50, no. 9
(November 2010): 1323–34, https://link.springer.com/
article/10.1007/s11340-010-9411-x#Abs1.

9   *250 to 325 million years ago*
In the *Dragonflies Q&A Guide,* Ann Cooper says 325 million.
Dragonflies and Damselflies: A Natural History says, "at least
250 million years," 9.

9   *with wingspans of up to two feet or more*
Sarah Zielinski, "14 Fun Facts about Dragonflies," *Smithsonian Magazine* (website), October 5, 2011, https://www
.smithsonianmag.com/science-nature/14-fun-facts-about
-dragonflies-96882693/.

9   *Scientists speculate that the more oxygen-enriched . . .*
Amanda Mascarelli, "Oxygen Boost Helps Dragonflies
Go Large," *Nature News Blog,* posted by Mark Peplow,
November 2, 2010, http://blogs.nature.com/news/2010/11/
oxygen_boost_helps_dragonflies.html.

Chapter 2

12   *"The starting point must . . ."*
Carl Linnaeus, quoted in William Leach, *Butterfly
People: An American Encounter with the Beauty of the World*
(New York: Vintage Books, 2013), xx.

13   *In fact, they are not "true bugs,"*
For a great explanation of what a true bug is, see Chris
Helzer, "What Kind of Bug Is a Bug?" *Prairie Ecologist*
(blog), February 11, 2019, https://prairieecologist.com/
2019/02/11/what-kind-of-bug-is-a-bug/.

13   *"Head, thorax, abdomen . . ."*
I made a diligent effort to track down the author of the lyrics
to what is popularly known as "The Insect Song." There are
numerous variations and tunes found online and on You-
Tube, and my adaptation is my own, with thanks to Nancy
Lamia (who suggested I add the "one to ten") and LuAnne
Lewandowski, who helped enthusiastically sing them—both
of them members of the Northern Kane Wild Ones in
Illinois. I would welcome crediting the original insect song
lyricist in the next edition of the book, so if you know who it
is, please write me. Here's one version of it, slightly different,
on YouTube: Dr. Jean, "Insect's Body: Learn and Sing about
Bugs with Dr. Jean," YouTube video, 2:40, May 1, 2016,

https://www.youtube.com/watch?v=6pe_p5FXE2g. You'll
see my version has been adapted for dragonflies.

15 *hind wings . . . "different wings."*
    Dragonflies of the North Woods, 1.

15 *There is a last segment . . . don't really walk.*
    Marla Garrison, email, June 3, 2019. A good simple
    description of dragonfly legs can be found in *Dragonflies
    Q&A Guide*, 23–24.

15 *Ode experts believe the antennae are important in detecting
    air movement . . . body orientation*
    Dragonflies and Damselflies: A Natural History, 13.

16 *sensory pits on the antennae, . . . "smelling" and detecting prey.*
    Nsikan Akpan, "Dragonflies Lack 'Smell Center,' but
    Can Still Smell," ScienceMag.org, American Academy
    for the Advancement of Science, March 21, 2014, https://
    www.sciencemag.org/news/2014/03/dragonflies-lack-smell
    -center-can-still-smell. This online article describes research
    done by invertebrate biologist Manuela Rebora in Italy.
    See also *Dragonflies and Damselflies: A Natural History*, 13.

16 *Scientists at this writing . . .*
    Dragonflies and Damselflies: A Natural History, 13.

16 *at about knee height*
    Low in the air? Likely a damselfly, although Pennants
    (dragonflies in the genus *Celithemis*) also fly low and fluttery.
    Dragonflies tend to be chilled in the morning and hang out
    in the grass until they warm up, so sometimes when you
    brush through knee-high grass, you flush out dragonflies.
    But most of the time, a knee-high flier will be a damselfly.

16 *damselfly . . . "young mistress."*
    Damselflies of Minnesota, Wisconsin, and Michigan, 1. I'm also
    indebted to its author, Robert DuBois, for his instructional
    information on the differences between dragonflies and
    damselflies, which helped me formulate some of the infor-
    mation on damselflies in this chapter, including
    "hammerhead" eyes.

17 *and touch, or nearly touch. . . . Eyes close together?*
    Some information here was nuanced and helped by

comments from Marla Garrison in her June 3, 2019, email and Dennis Paulson in his November 20, 2019, email.

17 *up to thirty thousand visual units honeycombed together*
Dragonflies Q&A Guide, 17.

17 *difficult to sneak up on a dragonfly.*
A great description of trying to net a dragonfly is found in *Dragonflies of the North Woods*, 18.

17 *five thousand to ten thousand individual light-sensitive units*
Damselflies of Minnesota, Wisconsin, and Michigan, 2.

17 *a dragonfly's color vision . . . polarized light*
Dragonflies and Damselflies: A Natural History, 54–55.

17 *Three other "simple eyes" . . .*
For more information on this see *Dragonflies Q&A Guide*, 18, and *Dragonflies and Damelflies: A Natural History*, 13. *Dragonflies of the North Woods* has some good detailed drawings of the eyes and eye area, viii.

18 *"Me—I've gotta be me!"*
"I've Gotta Be Me," lyrics by Walter Marks, originally written in 1967 for the Broadway musical *Golden Rainbow* but made a hit by Sammy Davis Jr. on an album of the same title, Reprise Records, 1968.

18 *jump into flight . . .*
Email conversation between Dennis Paulson and the author, November 20, 2019.

Chapter 3

22 *"Change is inevitable . . ."*
Robert C. Gallagher, "Quotes," Goodreads, accessed October 25, 2019, https://www.goodreads.com/quotes/78574-change-is-inevitable—except-from-a-vending-machine.

23 *Wisconsin's Horicon Marsh*
Discover more about Horicon Marsh via its website, https://www.horiconmarsh.org/.

24 *hatch in as short a period as five days . . .*
Dragonflies and Damselflies: A Natural History, 132.

24 *overwintering in some species*
"Life Cycle and Biology," British Dragonfly Society,

accessed October 25, 2019, https://www.british
-dragonflies.org.uk/content/biology-ecology.

24 *butterflies' more highly evolved "complete metamorphosis"*
For more on this topic, see Ferris Jabr, "How Did Insect
Metamorphosis Evolve?" Scientific American (website),
August 10, 2012, https://www.scientificamerican.com/
article/insect-metamorphosis-evolution/.

25 *A nymph moves through the water . . . its branchial basket*
Garrison, email, June 3, 2019. Thank you for the
wordsmithing, Marla, that made this chapter stronger.
See also "Introduction to the Odonata."

25 *"The Miraculous Rectum"*
Argia 30, no.1, page 31; used by permission of
Dennis Paulson.

26 *jaws comparable to a spring-loaded hinge*
Mead describes a "distendable" jaw (labium); see
Dragonflies of the North Woods, 7. Or "distensible" hinge,
which is what I used here, as per Paulson (email
conversation, November 20, 2019).

26 *eleven to fifteen times*
"Introduction to the Odonata."

26 *mayfly, which is said to molt more than fifty times.*
Justin W. Leonard, "Mayfly," in *Encyclopaedia Britannica
Online*, accessed October 25, 2019, https://www.britannica
.com/animal/mayfly.

26 *Information on molting*
Teale's accessible writing did a lot to help me understand
molting. I also learned a lot through an email conversation
with Marla Garrison, June 3, 2019. See Teale, *Strange Lives
of Familiar Insects*, and "Arthropod Molting: Why and How
It Happens," Kaleigh Jaron, Micscape (website), accessed
October 25, 2019, https://www.microscopy-uk.org.uk/
mag/artnov14macro/Kaleigh%20Jaron%20Arthropod
%20Molts.pdf.

26 *The new cuticle . . .*
Conversation with Marla Garrison, June 13, 2019. Her
excellent words.

26   *lenses of the eyes . . .*
     Teale, *Strange Lives of Familiar Insects*, 78.

26   *several months . . . seven years.*
     Marla Garrison notes "several months to several years"
     (email, June 3, 2019); Mead says "from as little as
     four weeks . . . to eight years" (*Dragonflies of the North
     Woods*, 7). As you can see, there is a lot of mystery in
     Odonate biology!

26   *Altitude and latitude*
     "Introduction to the Odonata."

26   *water temperature and the length of a growing season*
     Dragonflies of the North Woods, 7.

26   *Observes Edwin Way Teale, "Such changes . . . creature of the air."*
     Teale, *Strange Lives of Familiar Insects*, 78.

27   *stopped growing . . .*
     Teale, *Strange Lives of Familiar Insects*, 78.

27   *at Busse Woods . . .*
     "Busse Woods," Forest Preserves of Cook County,
     accessed October 25, 2019, https://fpdcc.com/places/
     locations/busse-woods/.

27   *"If you fear change . . ."*
     Robin Wall Kimmerer, *Gathering Moss: A Natural and
     Cultural History of Mosses* (Corvallis: Oregon State
     University Press, 2003), 35.

28   *Mead says that as many as 90 percent . . .*
     Check out more on the natural history of the dragonfly
     from Kurt Mead in *Dragonflies of the North Woods*, 1–11,
     https://www.mndragonfly.org/html/life-cycle.html.

28   *"At this point, ... ceased to be a water-breathing creature."*
     Teale, *Strange Lives of Familiar Insects*, 78.

28   *just before dawn*
     Garrison, email, June 3, 2019. Marla advised that I
     use "just before dawn" here. You'll read other sources
     that say night. See, for example, Chris Earley, *Dragonflies:
     Catching, Identifying, How and Where They Live* (Richmond
     Hill, Ont.: Firefly Books, 2013), 7. We have much to
     learn about dragonflies!

29   *briefly backward*
     Conversation with Dennis Paulson, November 20, 2019.

29   *Dragonfly populations of various species . . . long-term survivors.*
     Good information here on habitat impacts comes
     from a telephone conversation with Marla Garrison
     on April 12, 2019.

31   *"Since we go through this . . ."*
     Sigurd Olson, *Reflections from the North Country*
     (Minneapolis: University of Minnesota Press, 1998), 73.

31   *"Beauty and grace are performed . . ."*
     Annie Dillard, *Pilgrim at Tinker Creek* (New York:
     Harper Magazine Press, 1974), 8.

Chapter 4

34   *"Many of the reproductive habits . . ."*
     Teale, *Strange Lives of Familiar Insects*, 25.

35   *Dragonfly mating.*
     It would be difficult for me to understate the specific,
     accessible, and helpful information I gained from and the
     debt I owe to Kurt Mead's *Dragonflies of the North Woods* for
     helping this naturalist understand dragonfly reproduction.
     Other books with helpful dragonfly and damselfly mating
     and reproduction information that I consulted included
     Robert DuBois's *Damselflies of Minnesota, Wisconsin, and
     Michigan*, Dennis Paulson's *Dragonflies and Damselflies:
     A Natural History*, and the Stokes group's *Beginner's Guide
     to Dragonflies*. Marla Garrison, author of *Damselflies of
     Chicagoland*, gave me invaluable feedback and saved me
     from several errors. All five sources mentioned here are
     cited in full in the "Favorite Dragonfly-Chasing Guides"
     list preceding the endnotes.

36   *These males are show-offs.*
     Dragonflies of the North Woods, 5.

36   *grabs her by the back of the head*
     Stokes says "behind the eyes" for dragonflies and "by
     the neck" for damselflies (*Beginner's Guide to Dragonflies
     and Damselflies*, 8).

36 *Damselfly mating is similar, but slightly different.*
In our conversation on November 20, 2019, Dennis
Paulson explained to me that damselflies clasp the
junction between the prothorax and pterothorax with
their paraprocts. Try saying that five times fast!

36 *fifteen seconds or as long as an hour or more, until fertilization is
complete.*
Dragonflies of the North Woods, 6.

37 *"the female will refuse to mate . . . for oviposition."*
Garrison, email, June 9, 2019, in which she gave me
terrific information here on the courtship antics of
some dragonflies and damselflies.

37 *"Called sexual death feigning . . ."*
Sandhya Sekar, "This Female Insect Fakes Her Own
Death to Avoid Sex," *National Geographic,* May 1, 2017,
https://news.nationalgeographic.com/2017/05/death
-dragonflies-switzerland-mating-sex/. See also Rhassim
Khelifah, "Faking Death to Avoid Male Coercion:
Extreme Sexual Conflict Resolution in a Dragonfly,"
April 24, 2017, *Ecology Journal,* https:// /doi.org/10.1002/
ecy.1781; Patricia Edmonds, "Why This Insect Fakes
Death to Avoid Sex," *National Geographic,* June 2019,
https://www.nationalgeographic.com/magazine/2019/06/
moorland-hawker-dragonfly-fakes-death-to-avoid-sex/.

37 *may lay several thousand eggs . . .*
Kimberly Malcom, *The Dragonfly Life Cycle Explained—
Sites at Penn State* (2016), https://sites.psu.edu › files ›
ENGL202C-Life-Cycle-of-a-Dragonfly-160ontf

38 *Some dragonflies go it alone . . . promising area.*
Dragonflies of the North Woods, 6.

38 *Laying eggs in . . . strategy to avoid aggressive males.*
Sekar, "Female Insect Fakes Her Own Death."

38 *Most eggs will hatch in as little as five days or up to eight weeks . . .*
Dragonflies and Damselflies: A Natural History, 132.

38 *following spring*
"Life Cycle and Biology," British Dragonfly Society,
accessed October 25, 2019.

Chapter 5

I'm indebted in this chapter to several sources for my information, including: *Migratory Dragonfly Partnership*, n.d., http://www .migratorydragonflypartnership.org/uploads/_ROOT/File/ MDP-fact_sheet.pdf; "A Dragonfly ($\delta 2H$) Hisoscape for North America: A New Tool for Determining Natal Origins of Migratory Aquatic Emergent Insects," Keith A. Hobson, David X. Soto, Dennis R. Paulson, Leonard I Wassenaar, and John H. Matthews, *Methods in Ecology and Evolution*, 2012, https://besjournals .onlinelibrary.wiley.com/doi/full/10.1111/j.2041-210X.2012.00202.x; Kurt Mead's *Dragonflies of the North Woods*; and Dennis Paulson's *Dragonflies and Damselflies: A Natural History*.

42 *"Not all those who wander are lost."*
    J. R. R. Tolkien, *The Fellowship of the Ring*, quoted in
    "Journey Quotes," Goodreads, accessed October 25, 2019,
    https://www.goodreads.com/quotes/tag/journey.

43 *I'm at Kankakee Sands . . .*
    To discover more about Kankakee Sands, see "Places
    We Protect: Efroymson Restoration at Kankakee Sands,
    Indiana," Nature Conservancy, accessed October 25,
    2019, https://www.nature.org/en-us/get-involved/how-to
    -help/places-we-protect/kankakee-sands/.

43 *But did you know that several species of dragonflies also migrate?*
    "A Dragonfly ($\delta 2H$) Hisoscape for North America: A
    New Tool for Determining Natal Origins of Migratory
    Aquatic Emergent Insects," Keith A. Hobson, David X.
    Soto, Dennis R. Paulson, Leonard I Wassenaar, and John
    H. Matthews, *Methods in Ecology and Evolution*, 2012, https://
    besjournals.onlinelibrary.wiley.com/doi/full/10.1111/j.2041
    -210X.2012.00202.x.

44 *will travel at least as far as the Gulf Coast, Mexico, and
    the Caribbean.*
    Dragonflies of the North Woods, 12.

44 *According to Michele Blackburn . . .*
    Michele Blackburn, "Using Technology and Citizen
    Science to Understand Dragonfly Migration," *Xerces Blog*,
    April 27, 2016, https://xerces.org/blog/using-technology
    -and-citizen-science-to-understand-dragonfly-migration.

45   *"April is the cruellest month"*
T. S. Eliot, *The Waste Land*, in *T. S. Eliot: The Complete Poems and Plays, 1909–1950* (New York: Harcourt, Brace and World, 1971), 37.

46   *Since the 1880s . . . still poorly understood.*
"Migratory Dragonfly Monitoring," https://www
.citizenscience.gov/catalog/60/#.

46   *except Antarctica.*
Migratory Dragonfly Partnership, 1.

46   *at least five regular migrators*
These are all species that we believe migrate in Illinois.
See "Observations: 2019," Migratory Dragonfly
Partnership, accessed October 28, 2019, http://www
.migratorydragonflypartnership.org/observation/list.

46   *as many as eighteen migratory species in North America.*
Migratory Dragonfly Partnership, 1.

46   *move with weather fronts*
Marla Garrison, email, June 9, 2019.

46   *the answer is thought to be no, Dennis Paulson . . . tells me.*
Correspondence with the author, November 20, 2019.

46   *Damselflies may be dispersed some distances, however, carried by wind.*
Dragonflies and Damselflies: A Natural History, 140

47   *"Each day, tremendous clouds . . . sounding like sleet on dead leaves."*
Scott Weidensaul, *Living on the Wind* (New York: North Point Press; Farrar, Straus, and Giroux, 1999), 122.

47   *"That such delicate creatures undertake these epic journeys defies belief."*
Weidensaul, *Living on the Wind*, page x.

47   *Some dragonflies may travel over oceans . . .*
Dragonflies and Damselflies: A Natural History, 140.

47   *Studies, including one by Dr. Jessica Ware . . .*
"Dragonflies Migrate Further Than We Ever Guessed,"
IFLScience! accessed October 28, 2019, https://www
.iflscience.com/plants-and-animals/dragonflies-migrate
-further-we-every-guessed/.

47   *tiny transmitters . . .*
Dragonflies of the North Woods, 13. For more information on
the impracticalities of monitoring dragonflies by marking
and recapture, or transmitters, see the informative article
by Keith A. Hobson, David X. Soto, Dennis R. Paulson,
Leonard I Wassenaar, and John H. Matthews referenced
at the beginning of the notes to this chapter.

48   *American kestrels and merlins . . .*
Dragonflies of the North Woods, 12.

48   *Mississippi kites . . .*
Paulson, email conversation with the author,
November 20, 2019.

48   *"50 dragonflies for every three feet of his advance."*
Teale, *Strange Lives of Familiar Insects*, 82.

48   *temperature is said to play a role . . .*
Dragonflies and Damselflies: A Natural History, 141.

48   *Paulson tells me . . .*
Email correspondence with the author, November 20, 2019.

48   *Mead reminds us . . .*
Dragonflies of the North Woods, 12.

48   *Dragonflies will also disperse . . .*
Dragonflies and Damselflies: A Natural History, 140.

49   *"I am . . . looking for signs."*
Barbara Hurd, *Stirring the Mud: On Swamps, Bogs, and the
Human Imagination* (Boston: Beacon Press, 2001), 135–36.

Chapter 6

52   *"Diversity creates . . ."*
Barry Lopez, *Horizon* (New York: Alfred Knopf, 2019), 84.

53   *"So many books. So little time!"*
Often attributed to Frank Zappa, but well may be
apocryphal.

53   *Calico Pennant . . . yellow.*
Dragonflies of the North Woods, 216.

54   *Red Damsel (Amphiagrion)*
Note there is no species listed here, only the genus.
I received much helpful information on this genus from

dragonfly expert Marla Garrison, quoted here: "Red Damsel (*Amphiagrion*) is a species complex in controversy at the moment and Eastern Red Damsel is undoubtedly not what you are seeing based on the latest evidence— but rather Western Red Damsel or an intergradation or hybridization situation that has not yet been resolved through genetics and morphometrics. The most recent indication is that the Illinois *Amphiagrion* are actually 'abbreviatum' not 'saucium' based on some preliminary genetics by Jerrell Daigle and morphological studies by Ken Tennessen." The jury is still out. Another dragonfly mystery remains to be solved.

55 *"I love the names"*
Cody Considine, informal talk, 2017.

55 *3,109 dragonfly species and 3,212 damselfly species known to us worldwide*
Dragonflies and Damselflies: A Natural History, 200.
Information from the book has been updated to reflect the most up-to-date figures, per an email with Dennis Paulson on November 20, 2019.

55 *butterflies . . . about 20,000 species*
North American Butterfly Association, "Butterfly Questions and Answers," accessed October 29, 2019, https://www.naba.org/qanda.html.

55 *birds . . . just under 10,000 species*
Cornell Lab of Ornithology, "Clements Checklist," accessed October 29, 2019, http://www.birds.cornell.edu/clementschecklist/about/.

55 *"Our knowledge of the diversity . . . much to learn!"*
Dragonflies and Damselflies: A Natural History, 200.

55 *Costa Rica . . .*
Natali Anderson, "Beautiful New Species of Dragonfly Discovered," *SciNews*, November 6, 2019, http://www.sci-news.com/biology/gynacantha-vargasi-07774.html.

56 *North America . . . 469 Ode species*
Dragonflies Q&A Guide, 10; cited by Paulson and Sidney Dunkle in *A Checklist of North American Odonates*, 2020.

56 *Illinois has just under 150 distinct species*
Odonata Central, "Choose a Region: Illinois," accessed
October 29, 2019, https://www.odonatacentral.org/
index.php/ChecklistAction.showChecklist/location_id/35.
Keep in mind this number will stay pretty flexible
(Garrison, email, June 3, 2019).

57 *"Curiosity, imagination . . . atrophy with neglect."*
Paul Gruchow, *Boundary Waters: Grace of the Wild*
(Minneapolis: Milkweed Editions, 1997), 131. Gruchow
(1947–2004) wrote compellingly about the rural life
and the natural world in his books. If you haven't
read him before, try his books *Grass Roots: The Universe
of Home* and *Journal of a Prairie Year.*

Chapter 7

60 *"Dragonflies were as common as sunlight . . ."*
W. S. Merwin, "After the Dragonflies." See Carol
Rumens, "Poem of the Week: After the Dragonflies by
WS Merwin," *The Guardian*, October 10, 2016, https://
www.theguardian.com/books/booksblog/2016/oct/10/
poem-of-the-week-after-the-dragonflies-by-ws-merwin.

62 *"Nature is a spendthrift only of what she has the most."*
John Burroughs, *The Complete Nature Writings of John
Burroughs*, vol. 6, *Riverby* (New York: William H. Wise,
1904), 195–196.

62 *Insects may make up 80 percent of world's species . . . from two to
thirty million species of insects in the world.*
Department of Systematic Biology, Entomology Section,
National Museum of Natural History, *BugInfo: Numbers
of Insects (Species and Individuals)*, Information Sheet
no. 18, Smithsonian Institution, accessed October 30,
2019, https://www.si.edu/spotlight/buginfo/bugnos.

64 *Praying mantises . . . consume hummingbirds.*
Dillard, *Pilgrim at Tinker Creek*, 55. Also, email conversation
with Dennis Paulson, November 20, 2019.

64 *Teale spins the story of a two-pound trout . . . unwary smaller species.*
Teale, *Strange Lives of Familiar Insects*, 82.

64 *"Every live thing is a survivor . . ."*
Dillard, *Pilgrim at Tinker Creek*, 6–7.

64 *"bloodthirsty ogres, stalking endlessly for living prey."*
Teale, *Strange Lives of Familiar Insects*, 74.

65 *"It's a hell of a way to run a railroad."*
Dillard, *Pilgrim at Tinker Creek*, 176

65 *The pressure . . . reducing life there to a universal chomp."*
Dillard, *Pilgrim at Tinker Creek*, 168.

65 *more than sixty-five thousand women . . . twelve thousand were*
projected to die from it in 2019.
"Uterine Cancer: Statistics," Cancer.Net, January 2019,
https://www.cancer.net/cancer-types/uterine-cancer/
statistics. One early warning symptom is post-
menopausal spotting.

66 *"Here is the world. . . . Don't be afraid."*
Frederick Buechner, "September 9, 2016: Grace,"
Frederick Buechner (website), September 9, 2016,
https://www.frederickbuechner.com/quote-of-the-day/
2016/9/9/grace.

66 *I find a dragonfly lifeless on the ground.*
I first wrote about this incident in *By Willoway Brook:*
Exploring the Landscape of Prayer (Orleans, Mass.:
Paraclete Press, 2004), 33.

67 *"nature loves the idea . . . pizzazz."*
Dillard, *Pilgrim at Tinker Creek*, 180.

67 *"I've found . . . we're privileged to live on."*
Kim Smith, email, April 9, 2019.

68 *I saw my first Ashy Clubtail*
An Ashy Clubtail was ID'd by the author, but does not
appear on the park's Odonate inventory.

69 *"Joy as I see it . . ."*
Susan Goldsmith Wooldridge, *Poemcrazy: Freeing Your Life*
with Words (New York: Three Rivers Press, 1997), 148.

Chapter 8

72   *"In human culture is the preservation of wildness."*
Wendell Berry, "Weeds Are Us," *New York Times Magazine*,
November 5, 1989, https://michaelpollan.com/articles
-archive/weeds-are-us/.

74   *samurai helmet with a dragonfly on top*
Many dragonfly-themed samurai helmets are in
museums; for one example, see Metropolitan Museum
of Art, "Helmet (*Kawari-kabuto*) Surmounted by a
Dragonfly, 18th Century; Restorations, 2015," The
Met (website), accessed October 30, 2019, https://www
.metmuseum.org/art/collection/search/27591.
Thanks to Andrés Ortega for sharing this form of
art with me through one of his dragonfly workshops.

74   *smooth jazz album*
"Dragonfly Summer," track 3 on Michael Franks,
Dragonfly Summer, Reprise, 1993, compact disc.

74   *Early in Fleetwood Mac's career . . .*
Note that Stevie Nicks was not with the band when this
song was released.

74   *William Henry Davies, a little-known Welsh poet.*
See "The Dragonfly" by W. H. Davies, accessed
October 30, 2019, https://hort330-deshields.weebly
.com/poem.html.

75   *"It was a fleeting visit all too brief . . ."*
"Dragonfly," MP3 audio, on Blake, *The First Snow*,
released December 24, 2011, Bandcamp, https://
thisisblake.bandcamp.com/track/dragonfly.

75   *Alfred Lord Tennyson's poem "The Dragon-Fly"*
Tennyson, *Poetic and Dramatic Works*, "The Two Voices"
(1883), found in *The Poetic and Dramatic Works of Alfred
Lord Tennyson* (New York: Houghton Mifflin, 1898), 30.
I was first introduced to this poem in the book *A Dazzle
of Dragonflies* by Forrest L. Mitchell and James L. Laswell
(College Station: Texas A&M University Press, 2005), 29.

75   *"As kingfishers catch fire, dragonflies draw flame."*
Gerard Manley Hopkins, "As Kingfishers Catch Fire,"

Poetry Foundation, https://www.poetryfoundation.org/
poems/44389/as-kingfishers-catch-fire.

75 *poem by Vachel Lindsay from the 1920s, "The Dragon-Fly Guide"*
Vachel Lindsay, "The Dragon-Fly Guide," *Poetry*,
April 1926, 3, https://www.poetryfoundation.org/
poetrymagazine/browse?contentId=17132.

75 *Throughout Japanese history . . . the island nation.*
Ron Lyons, "Cultural Odonatology References"
(website), accessed October 30, 2019, http://casswww
.ucsd.edu/archive/personal/ron/CVNC/odonata/cultural
_odonatology.html.

75 *Akitsu-shima, "Island of the Dragonfly" . . . make them wealthy.*
Dazzle of Dragonflies, 30–31.

75 *"Crimson pepper pod . . . darting dragonfly"*
Basho, cited in Cindy Crosby, "Prairie Dragonfly
Mysteries," *Tuesdays in the Tallgrass* (blog), March 13, 2018,
https://tuesdaysinthetallgrass.wordpress.com/2018/03/13/
prairie-dragonfly-mysteries/.

75 *aptly named collection* Dragonfly Haiku
Kobayashi Issa, Ken Tennessen, and Scott King, *Dragonfly
Haiku* (Northfield, Minn.: Red Dragonfly Press, 2016).

76 *"Maybe you've seen a dragon-fly . . . pitiful to see."*
Precious Bane, Mary Webb (New York: Modern Library/
Random House, 1926). To read more about Webb, see
Eve M. Kahn, "Shining a Light on a Forgotten Poet,"
New York Times, January 28, 2010, https://www.nytimes
.com/2010/01/29/arts/design/29antiques.

76 *In Bali, . . . use in soup.*
Vivienne Kruger, "Food of the Gods 2007: Traditional
Village Food; Cooking in the Compound, Part 1,"
Bali Advertiser, accessed October 30, 2019, https://
www.baliadvertiser.biz/food_1/.

76 *dragonfly larvae on skewers for sale in a Chinese market.*
Paulson has a great photo in his book of dragonfly larvae
as culinary treats for sale; see *Dragonflies and Damselflies:
A Natural History*, 167.

76 *"reflect the sun in tiny diamond facets."*
    Karen Johnson, email, April 12, 2019. Her work is
    at www.karensnatureart.com.

76 *Arthur Pearson*
    Find Arthur Melville Pearson's work at www
    .arthurmelvillepearson.com.

76 *Peggy Macnamara*
    Find the amazing Peggy Macnamara's work at www
    .peggymacnamara.com.

77 *wife "caught the bug" . . . has begun a dragonfly quilt.*
    Mark Jordan, email, April 10, 2019.

77 *"Everything about them is challenging. . . . It is spiritual."*
    Saksena, email, April 12, 2019.

77 *"No better way . . . not throw the customary stone."*
    "Oscar Wilde Quotes," BrainyQuote, accessed
    October 30, 2019, https://www.brainyquote
    .com/quotes/oscar_wilde_700618.

Chapter 9

Part of this chapter first appeared in a slightly different format in
Cindy Crosby, "Under the Prairie Ice," *Tuesdays in the Tallgrass*
(blog), January 29, 2019, https://tuesdaysinthetallgrass.wordpress
.com/tag/andres-ortega/.

80 *"The last word in ignorance . . ."*
    Aldo Leopold, "Conservation," in *Round River: From the
    Journals of Aldo Leopold*, ed. Luna B. Leopold (Oxford:
    Oxford University Press, 1966), 146–47. He wrote this
    piece in 1938, yet it still rings true eighty-plus years later.

81 *at Blackwell Forest Preserve*
    Discover more about the Blackwell Forest Preserve
    via the Forest Preserve District of DuPage County
    website, https://www.dupageforest.org/places-to-go/
    forest-preserves/blackwell.

81 *Forest Preserve of DuPage County ecologist Andrés Ortega*
    Andrés Ortega, interview by the author, January 15, 2019;
    Andrés Ortega, email, May 29, 2019.

81 *"seem to be rare and localized"*
Everett D. Cashatt, "Hine's Emerald Dragonfly,"
Illinois State Museum, last updated October 4, 2012,
https://www.museum.state.il.us/research/entomology/
hines/mainpage.html.

81 *only dragonfly on the Federal Endangered Species List*
Cashatt, "Hine's Emerald Dragonfly"; U.S. Fish
and Wildlife Service Midwest Region, "Midwest
Region Endangered Species: Hine's Emerald Dragonfly
(*Somatochlora hineana*)," last updated May 29, 2019, https://
www.fws.gov/midwest/endangered/insects/hed/index.html.

81 *Urban Stream Research Center*
Discover more about the Urban Stream Research
Center via the Forest Preserve District of DuPage County
website, https://www.dupageforest.org/plants-wildlife/
restore-conserve/habitats/urban-stream-research-center.

82 *lays its eggs . . . devil crayfish.*
Timothy E. Vogt and Everett D. Cashatt, *Survey Site
Identification for Hine's Emerald Dragonfly* (Somatochlora
hineana) *in Illinois: Final Report*, March 2007, http://
www.museum.state.il.us/research/entomology/hines/
2007_IL_HED_final_report.pdf.

83 *lays more than five hundred eggs during her lifetime.*
Dragonflies and Damselflies: A Natural History, 190.

83 *Hine's Emerald description*
Dragonflies of the North Woods, 166–67.

84 *"darker tinting . . . as they become older."*
Cashatt, "Hine's Emerald Dragonfly."

84 *first observed in 1931 . . . twenty different sites.*
Cashatt, "Hine's Emerald Dragonfly"; U.S. Fish and Wild-
life Service Midwest Region, "Midwest Region Endangered
Species: Hine's Emerald Dragonfly"; Michigan Natural
Features Inventory, MSU Extension, "Plants and Animals:
Somatochlora hineana, Hine's Emerald Dragonfly," Michigan
State University, accessed November 1, 2019, https://
mnfi.anr.msu.edu/species/description/12124; Candace
Davis, "MDC: Federally Endangered Hine's Emerald

Dragonfly Found on New Site," Missouri Department of Conservation, June 15, 2015, https://mdc.mo.gov/newsroom/mdc-federally-endangered-hine-s-emerald-dragonfly-found-new-site.

84 *declared endangered . . . federal endangered status in 1995.*
Cashatt, "Hine's Emerald Dragonfly."

84 *bitter battle over drilling wells (which would disrupt the Hine's Emerald dragonfly habitat)*
For more on this battle, see the following newspaper accounts: Elizabeth Birge, "Rare Dragonfly May Bar Way of I-355 Extension," *Chicago Tribune,* July 4, 1993, https://www.chicagotribune.com/news/ct-xpm-1993-07-04-9307040021-story.html; Frank Vaisvilas, "Endangered Dragonfly Presenting Problem in Lockport," Daily Southtown, *Chicago Tribune,* July 31, 2015, https://www.chicagotribune.com/suburbs/daily-southtown/news/ct-sta-lockport-dragonfly-problem-st-0802-20150731-story.html; "I-355 Construction 'Dragon' Because of Endangered Fly," *Daily Journal* (Kankakee), January 15, 2005, https://www.daily-journal.com/news/local/i–construction-dragon-because-of-endangered-fly/article_7cf65aa8-c59d-50dd-b893-a6bab18581b8.html.

85 *habitat destruction may be one of the biggest threats to recovery.*
Dragonflies of the North Woods, 166. See also Vogt and Cashatt, *Survey Site Identification for Hine's Emerald Dragonfly*; Cashatt, "Hine's Emerald Dragonfly."

85 *threatened by vehicular traffic*
Dragonflies and Damselflies: A Natural History, 190.

85 *climate change*
Email with Dennis Paulson, November 20, 2019.

85 *between ten million and fifty million species of plants and animals*
U.S. Fish and Wildlife Service, *Why Save Endangered Species?* July 2005, https://www.fws.gov/nativeamerican/pdf/why-save-endangered-species.pdf.

85 *"To keep every cog and wheel is the first precaution of intelligent tinkering"*
Leopold, "Conservation."

86 *"are of . . . value to the Nation and its people."*
Quoted in U.S. Fish and Wildlife Service, *Why Save Endangered Species?*

86 *"No one knows . . . chain reaction affecting many others."*
U.S. Fish and Wildlife Service, *Why Save Endangered Species?*

86 *"insect apocalypse"*
Jarvis, "Insect Apocalypse." The phrase "insect apocalypse" has been widely used since the 2018 publication of Jarvis's article.

86 *"Distraction and indifference . . . harrowing to face."*
Lopez, *Horizon*, 47.

86 *federally threatened eastern prairie fringed orchid*
U.S. Fish and Wildlife Service Midwest Region, "Midwest Region Endangered Species: Eastern Prairie Fringed Orchid (*Platanthera leucophaea*)," last updated October 3, 2019, https://www.fws.gov/midwest/Endangered/plants/epfo/index.html.

87 *"they took their light with them when they went."*
Merwin, "After the Dragonflies."

Chapter 10

90 *"Stuff your eyes with wonder."*
Ray Bradbury, *Fahrenheit 451* (New York: Simon and Schuster, 2013, 150).

92 *"all natural objects . . . open to their influence."*
Ralph Waldo Emerson, *The Complete Works of Ralph Waldo Emerson*, vol. 1, *Nature Addresses and Lectures*, https://quod.lib.umich.edu/e/emerson/4957107.0001.001/63/63/1?node=4957107.0001.001:11&view=text.

92 *"In our eyes . . . the burden of our metaphors."*
Michael Pollan, *Second Nature: A Gardener's Education* (New York: Atlantic Monthly Press, 1991), 95.

93 *One dragonfly monitor told me . . . flies and mosquitoes.*
Email interview with Karen Remkus, April 10, 2019.

93 *Zuni myth from the Pueblo people*
Lyons, "Cultural Odonatology References." According to this Odonate aficionado, the Zuni myth "The Boy

Who Made Dragonfly" was first recorded by the anthropologist Frank Hamilton Cushing in 1883. Based on Cushing's recording, the twentieth-century author Tony Hillerman published a version of the story in 1972 that has since been reissued; see, e.g., *The Boy Who Made Dragonfly: A Zuni Myth Retold by Tony Hillerman* (Albuquerque: University of New Mexico Press, 1986).

93   *"When I was growing up . . . never been stung by one."*
Curt Oien, email, April 17, 2019.

94   *"that 'devil's darning needles' were poisonous . . . "*
References to the devil and other names from *Dazzle of Dragonflies*, 209–212. Many thanks to the authors. This book was a great resource for lists of dragonfly nicknames and dragonfly lore that were useful in writing this chapter.

94   *"the devil's steelyard"*
There are numerous references online about the shape of the dragonfly's body looking like a weighing tool, from Swedish folklore. Supposedly, if the dragonfly flew around your head, it was weighing your soul (not a good sign). But when I tried to track down a definitive source for this story, I came up empty. If you have one, please contact me! I'll include it in the next edition.

95   *In Tahitian lore, . . . "It was a god . . . they were being robbed."*
Lyons, "Cultural Odonatology References," crediting Teuira Henry, "Ancient Tahiti" (ca. 1928), *Bernice Bishop Museum Bulletin* 48, 391.

95   *"King of Death" . . . "Dragonfly of the Ancestors"*
Dazzle of Dragonflies, 209–12.

95   *And in the Philippine Islands . . .*
Lyons, "Cultural Odonatology References," crediting *Cultural Dictionary of Philippine Folk Beliefs and Customs*, by R. Francisco and S. Demetrio (Pasay City: Xavier University Cagayun de Oro City, Modern Press, 1970), 201.

96   *"snakefeeder"*
B. E. Montgomery, "Why Snakefeeder? Why Dragonfly? Some Random Observations on Etymological Entomology," *Proceedings of the Indiana Academy of Science* 82 (1972): 235–41.

96 *needles, swords, arrows, and pins*
    Dazzle of Dragonflies, 210–11.

Chapter 11

100 *"You can observe a lot by watching."*
     Yogi Berra, *You Can Observe a Lot by Watching: What
     I've Learned about Teamwork from the Yankees and Life*
     (New York: John Wiley and Sons, 2009).

102 *nymphs face off . . . errant dragonfly.*
     Cashatt, "Hine's Emerald Dragonfly."

102 *it's important to track it over many years; even for generations*
     Dave Goulson quoted in Jarvis, "Insect Apocalypse
     Is Here," 43.

102 *"With so much abundance . . . and who would pay for it?"*
     Jarvis, "Insect Apocalypse Is Here," 43.

103 *Citizen science "is often seen . . . with the scientific objectives
     of scientists themselves"*
     J. Silverton, "A New Dawn for Citizen Science," *Trends
     in Ecology and Evolution* 24, no. 9 (2009): 467–71. Another
     journal article that inspired this chapter is Hauke Reisch
     and Clive Potter, "Citizen Science as Seen by Scientists:
     Methodological, Epistemological, and Ethical Dimensions,"
     Public Understanding of Science 23, no. 1 (2014), https://doi
     .org/10.1177/0963662513497324.

103 *monarch migration . . . efforts.*
     For more information, see the Monarch Joint Venture
     website, https://monarchjointventure.org/get-involved/
     study-monarchs-citizen-science-opportunities.

104 *The Morton Arboretum's one-hundred-acre Schulenberg
     Prairie . . .*
     For information on the Morton Arboretum in Lisle,
     Illinois, see its website, https://www.mortonarb.org/.

104 *Nachusa Grasslands . . .*
     For information on the Nature Conservancy's Nachusa
     Grasslands in Franklin Grove, Illinois, see its website,
     https://www.nachusagrasslands.org/.

108 *Ohio Dragonfly Survey*
A three-year project funded by the Ohio Department
of Natural Resources through the Ohio Biodiversity
Conservation Partnership; see SciStarter, "The Ohio
Dragonfly Survey," last updated September 6, 2019,
https://scistarter.org/the-ohio-dragonfly-survey.

108 *"Many areas here . . . old mining road."*
MaLisa Spring, email, April 9, 2019.

108 *"I also don't have . . . spring rainfalls."*
Smith, email, April 9, 2019.

109 *"No one has described . . . We hit pay dirt!"*
Kurt Mead, telephone interview by author, April 10, 2019.

109 *Mead told me . . . what the impact of the mining is on dragonflies
and damselflies.*
Kurt Mead, telephone interview by author, April 10, 2019.

111 *the meaning of* amateur . . . *"lover."*
Jarvis, "Insect Apocalypse Is Here."

111 *"If I was going to volunteer . . . much more to be done."*
Oien, email, April 17, 2019.

Chapter 12

114 *"Hanging out with great and kind people . . ."*
Mead, interview, April 9, 2019.

116 *For Ode expert Dennis Paulson . . .*
Paulson, email correspondence, December 4, 2019.

117 *"So, I started noticing . . . eyes are absolutely captivating."*
Saksena, email, April 12, 2019.

117 *"I saw kingbirds . . . dragonfly in the background of a bird photo
I took."*
Smith, email, April 9, 2019.

118 *Dragonfly Society of the Americas meeting*
For more information about the conference, check
the Dragonfly Society of the Americas website, https://
www.dragonflysocietyamericas.org/.

117 *blogs about dragonflies . . .*
Check out Kim's blog, *Nature is My Therapy*, https://
natureismytherapy.com.

117 *"a nice, crisp photo."*
Smith, email, April 9, 2019.

118 *"Once I told an old friend . . . understand each other."*
Mark Donnelly, email, April 15, 2019.

118 *"I wanted to get proficient . . . has been amazing."*
Mead, email, April 19, 2019.

119 *"It is nice to see so much excitement from others"*
Spring, email, April 10, 2019.

120 *"It is also really fun . . . make a big difference."*
Spring, email, April 9, 2019.

121 *"One day . . . happy dance on the dock!"*
Linda Gilbert, email, April 11, 2019.

121 *"Biology is life . . . has changed my life"*
Garrison, telephone conversation, April 12, 2019.

122 *"Many of us had looked . . . arrived at in cars!"*
Oien, email, April 17, 2019.

Chapter 13

126 *"What would the world be . . ."*
Gerard Manley Hopkins, "Inversnaid," in *Poems of Gerard Manley Hopkins*, ed. Robert Bridges (London: Humphrey Milford, 1918), 33, https://en.wikisource.org/wiki/Poems _of_Gerard_Manley_Hopkins/Inversnaid.

128 *"There was a dragonfly . . . staring into each other's souls."*
Adult student, from a class with the author, October 7, 2019.

128 *"tune in" to what we specifically want to see*
Alexandra Horowitz, *On Looking: Eleven Walks with Expert Eyes*, Center Point Large Print Edition (New York: Simon and Schuster, 2013), 22–26.

128 *quoting Susan Morse*
On Looking: Eleven Walks with Expert Eyes, 160.

130 *"butterfly binoculars"*
Dragonflies of the North Woods has a great list of close-focus binoculars (see page 266).

131 *"face blindness"*
Julie Leibach, "What Is Face Blindness?" *Science*

Friday, February 11, 2016, https://www.sciencefriday
.com/articles/what-is-face-blindness/.

132 *"Some species . . . bring them down."*
Teale, *Strange Lives of Familiar Insects*, 80.

132 *he recalls using a .22 caliber revolver . . .*
Email correspondence with Dennis Paulson,
November 20, 2019, and December 4, 2019. The
dragonfly, he noted, never flew within range of his net.

132 *stories about putting flour in a gun*
Garrison, email, June 9, 2019.

132 *"The Darners and Spiketails are not easy to photograph or net"*
Spring, email, April 10, 2012.

132 *"Nothing focuses the attention . . . touching the infinite."*
Jim Lemon, email, April 12, 2019.

133 *"The one thing that completely blows my mind . . . they continue
to miss it!"*
Saksena, email, April 12, 2019.

133 *"Habitat loss is heartbreaking to anyone with sentiment for nature"*
Lemon, email, April 12, 2019.

134 *old world globe . . . Here be dragons.*
https://www.theatlantic.com/technology/archive/2013/12/
no-old-maps-actually-say-here-be-dragons/282267/

Chapter 14

138 *"Most children have a bug period, and I never grew out of mine."*
Edward O. Wilson, *Naturalist* (Washington, D.C.:
Island Press, 1994), 53.

140 *"Someone in my family . . . stay away from it."*
Lemon, email, April 12, 2019.

140 *One of my favorites . . .*
Conversation with Gillian Crosby, summer 2018

141 *"A child's world . . . sources of our strength."*
Rachel Carson, *The Sense of Wonder* (Covelo, Calif.:
Nature Company at Yolla Bolly Press, 1990).

141 *"The art of seeing has to be learned"*
Marguerite Duras, "Eleven Walks," quoted in Maria

Popova, "The Art of Looking: Eleven Ways of Viewing the Multiple Realities of Our Everyday Wonderland," Brain Pickings, August 12, 2013, https://www.brainpickings.org/2013/08/12/on-looking-eleven-walks-with-expert-eyes/.

142   *"baby dragonflies are butt breathers and . . . swim with jet propulsion."*
Oien, email, April 17, 2019.

142   *"The excitement of the 'hunt' . . . breaking down barriers for kids who have learned to be afraid of bugs."*
Mead, email, April 9, 2019.

143   *"We almost always found dragonfly and damselfly nymphs."*
Karen Remkus, email, April 9, 2019.

144   *iNaturalist*
Check out this wonderful online resource at https://www.inaturalist.org.

144   *Dragonfly ID app*
"What Is the Dragonfly ID App?" Birdseye Apps, accessed November 5, 2019, http://www.birdseyebirding.com/apps/dragonfly-id-app/.

144   *"We teach our children . . . to wake up."*
Annie Dillard, as quoted in Russell, *Obsession with Butterflies*, 206.

145   *"Butterflies wake us up."*
Russell, *Obsession with Butterflies*, 206.

145   *"What is the value . . . strength that will endure as long as life lasts."*
Carson, *Sense of Wonder*.

Chapter 15

148   *"Anthropocentric . . . recognition of its mystery."*
Michael Pollan, *Second Nature: A Gardener's Education* (New York: Atlantic Monthly Press, 1991), 192.

149   *One thing I did . . . put in a pond.*
Some of the process of digging the pond appeared in another form and is detailed further in Cindy Crosby, *Waiting for Morning* (Grand Rapids, Mich.: Baker, 2001).

149   *JULIE*
      JULIE, "About Us," accessed November 5, 2019, https://
      www.illinois1call.com/about/.

150   *Although the surface ices over . . .*
      Some dragonfly eggs may survive freezing solid,
      others may not. Some nymphs may survive a short
      period of freezing, but why take a chance?

152   *"A garden . . . something nearer poetry."*
      Pollan, *Second Nature*, 244.

152   *dragonflies . . . "drink" . . . absorb dew.*
      Dragonflies of the North Woods, 11.

152   *Your water feature . . . a place to scramble around.*
      A good resource for backyard ponds that are Ode-
      friendly can be downloaded from Migratory Dragonfly
      Partnership; Celeste Mazzacano, Dennis Paulson, and
      John Abbott, *Backyard Ponds: Guidelines for Creating and
      Managing Habitat for Dragonflies and Damselflies* (Portland,
      Ore.: Migratory Dragonfly Partnership, 2014), http://www
      .migratorydragonflypartnership.org/uploads/_ROOT/File/
      Pond_Habitat_Guidelines_Odonates_Final_Websec.pdf.

153   *Dodds recommends siting your water feature in full sun.*
      Ruary MacKenzie Dodds, *The Dragonfly-Friendly Gardener:
      Create a Garden Home for Dragonflies and Damselflies*
      (Glasgow: Saraband, 2016), 16.

154   *A diversity of foliage . . . a place to perch.*
      Catherine Mabry and Connie Dettman, "Odonata Richness
      and Abundance in Relation to Vegetation Structure in
      Restored and Native Wetlands of the Prairie Pothole
      Region, USA," *Ecological Restoration* 28, no. 4 (November
      2010), 475–84.

155   *Did you know that mosquitoes and wasps are pollinators?*
      Zoe Statman-Weil, "*Aedes communis*: The Pollinating
      Mosquito," U.S. Forest Service (website), https://www
      .fs.fed.us/wildflowers/pollinators/pollinator-of-the-month/
      aedes_communis.shtml.

155   *they may benefit your garden by eating insect pests*
      University of Florida, "A New Angle on Flowers:

Fish Are Players in Pollination," news release, October 5, 2005, http://news.ufl.edu/archive/2005/10/a-new-angle-on-flowers-fish-are-players-in-pollination.html.

Chapter 16

158 *"Every one is inclined . . . same passion."*
Samuel Johnson, "No. 83: Tuesday, January 1, 1750," *The Rambler*, sec. 55–112, p. 168, in *The Collected Works of Samuel Johnson in Sixteen Volumes*, vol. 4, Electronic Text Center, University of Virginia Library, https://web.archive.org/web/20110222182726/http://etext.lib.virginia.edu/etcbin/toccer-new2?id=Joh4All.sgm&images=images/modeng&data=/texts/english/modeng/parsed&tag=public&part=29&division=div2.

159 *"Looking for Nature at the Mall."*
Jennifer Price, "Looking for Nature at the Mall," in *Uncommon Ground: Rethinking the Human Place in Nature*, ed. William Cronon (New York: W. W. Norton, 1995), 186–202.

159 *Nature Company retail stores*
Wikipedia, s.v. "The Nature Company," last modified October 6, 2019, https://en.wikipedia.org/wiki/The_Nature_Company.

160 *"We've filled our homes . . . or perhaps 'solitude.' "*
Price, "Looking for Nature at the Mall," 193, 194.

160 *"Why are we looking for nature with our credit cards?"*
Price, "Looking for Nature at the Mall," 197.

161 *"Is it possible . . . their only avenue to it is consumerism?"*
Price, "Looking for Nature at the Mall," 197, quoting from Pete Dunne, "In the Natural State," *New York Times*, May 7, 1989.

Chapter 17

This chapter first appeared in a slightly different form in an essay, "Blowing Bubbles: Much Ado about Nothing," which won the Paul Gruchow Memorial Prize for Writers Rising Up, 2013, copyright Cindy Crosby.

166 *"Pay attention. Be astonished. Tell about it."*
Mary Oliver, "Sometimes," in *Red Bird* (Boston:
Beacon Press, 2008), 37.

168 *"Tell me . . . precious life?"*
Mary Oliver, "The Summer Day," in *New and
Selected Poems* (Boston: Beacon Press, 1992), 94.

169 *Grandma's kitchen . . .*
Thoughts about my grandmothers appeared in a slightly
different form in *By Willoway Brook*, (Orleans, Mass.:
Paraclete Press, 2004), 57, 106.

170 *"Many of us will never see . . . suffice."*
Rick Kogan, "Audubon on the Avenue," *Chicago Tribune*,
September 30, 2007, https://www.chicagotribune.com/
news/ct-xpm-2007-09-30-0709270752-story.html.

170 *"New actions inevitably lead . . . notice how little we have
been noticing. . . ."*
Tristan Gooley, *How to Read Nature: Awaken Your Senses
to the Outdoors You've Never Noticed* (New York: Experiment
Publishing, 2017), 10.

173 *"To pay attention: This is our endless and proper work"*
Mary Oliver, "Yes! No!" in *Owls and Other Fantasies:
Poems and Essays* (Boston: Beacon Press, 2003), 27.

Epilogue

176 *"Let us go on, and take the adventure that shall fall to us."*
C. S. Lewis, *The Lion, the Witch, and the Wardrobe*
(1950; repr., New York: HarperCollins, 1978), 187.

177 *"The first dragonfly of the year"*
Scott King, "The First Dragonfly of the Year," in Issa,
Tennessen, and King, *Dragonfly Haiku*, 77.

178 *"Pay attention. Be astonished. Tell about it."*
Oliver, "Sometimes."

180 *"How we spend our days, of course, is how we spend our lives."*
Dillard, *The Writing Life (New York: Harper and Row, 1990).*

180 *"When it's over . . . married to amazement."*
Mary Oliver, "When Death Comes," in *New and
Selected Poems*, vol. 1 (Boston: Beacon Press, 1992), 10.

Closing Epigraph

181 *"Spring comes—"*
Ken Tennessen, "Spring Comes," in Issa, Tennessen,
and King, *Dragonfly Haiku,* 57.

ACKNOWLEDGMENTS

I owe many people a tremendous debt of gratitude for their assistance. The inimitable Dennis Paulson, author of *Dragonflies and Damselflies: A Natural History*, was kind enough to read the manuscript and give me detailed feedback and corrections, as well as encourage a dragonfly chaser he'd never met. What a gift! I can't overestimate the difference he made to this book. Thank you, Dennis. Grateful thanks to Ode expert Marla Garrison, author of *Damselflies of Chicagoland*, who read this manuscript in the early rough stages and gave me so much helpful, specific feedback. She was also a kind and enthusiastic encouragement. Thank you for your generosity, Marla. Kurt Mead, author of *Dragonflies of the North Woods*, shared his abundant knowledge and stories with me by email and by phone, and I can't thank him enough for his kindness. Many thanks to Forest Preserve District of DuPage County's Ecologist Andrés Ortega for his tour of the Urban Stream Research Center, his patient answers to all my questions, his terrific work with the Hine's Emerald dragonflies, and his reading of and suggested edits of the Hine's Emerald portion of the book. I'm grateful to Dr. Tim Cashatt and Dr. Timothy Vogt, who patiently took time with me at that long-ago Dragonfly Society of the Americas meeting in Decorah, Iowa, and helped me net my first dragonfly, an Autumn Meadowhawk. I'll never forget that experience. (See what you did?) Any errors remaining in this manuscript are my own.

Grateful, grateful thanks to my wonderful editorial director Jane Bunker, who believed in this book from the start; terrific marketing and sales director and fellow gardener JD Wilson, who has championed my writing from the beginning; awesome managing editor and manager of design and production Anne Gendler; creative director extraordinaire Marianne Jankowski; super marketing coordinator Anne Strother; and the amazing team at Northwestern University Press: Parneshia Jones, Trevor Perri, Patrick Samuel, Liz Alexander, Maia Rigas, Morris (Dino) Robinson, Emily Dalton, Amy Schultz, Courtney Smotherman, Liz Hamilton, and Laura Ferdinand. Tremendous thanks go to Lori Meek Schuldt for her copyediting of this manuscript. Grateful appreciation to Mary Klein for her excellent proofing skills. Thank you, team, for your guidance, your enthusiasm, and your tenacity in bringing this book to publication. I'm so delighted to be a part of your work.

Margaret "Peggy" Macnamara is one of the finest artists I now have the privilege to know. I'm grateful for her work here and for her passion for natural history, bringing the smallest creatures to the attention of so many. Thank you, Peggy, for being a partner on this journey and for being a kind and generous person.

The Morton Arboretum and Nachusa Grasslands have kindly allowed me to carry out my obsession with dragonflies and recruit others to monitor, and they have given me the support and encouragement to do my teaching work and volunteer work there. Thank you, Mark McKinney, Kurt Dreisilker, Kristin Sabatino, Spencer Campbell, and their staff who support the dragonfly monitoring efforts in Natural Resources. The Morton Arboretum's education department staff—Brooke Pudar, Megan Dunning—are enthusiastic in their support of the dragonfly and damselfly ID classes I teach each summer and have given me many opportunities to share my writing. Thank you. Kevin Earll, book buyer extraordinaire at the Morton Arboretum Store, and the staff have been tireless supporters of my books. Thank you.

Elizabeth Bach, Bill and Susan Kleiman, and Cody Considine have all been tremendously supportive of my seven-year dragonfly monitoring program at Nachusa Grasslands and helped me recruit and train monitors there. The Forest Preserve District of DuPage County and then-ecologist Tom Velat was one of the first to train me in learning the particulars of what a dragonfly is and how to stalk it in the field and report data. I'm grateful to these people and institutions for what they have taught me and how they introduce others to the magic of dragonflies.

Grateful thanks to my monitors, who spend countless hours in the field and make things happen for the Morton Arboretum and Nachusa Grasslands each season. I'm grateful for your enthusiasm and joy. Go Team Dragonfly!

A big shout-out to the Peggy Notebaert Nature Museum and Lalainya Goldsberry, Lily Bajas, and Allison Sacerdote-Velat, who work tirelessly on behalf of the natural world, and who at different times have helped my monitors and me with our data for Illinois dragonfly reporting. Their patience, participation, and love of the work is infectious. Thanks to Mary Samerdyke, Jeff Pines, and Paul and Rachel Samerdyke for the Horicon Marsh adventure. Grateful.

The staff at Peet's Coffee in Naperville, Illinois, and at Labriola Café in Oak Brook, Illinois, kept my coffee cup full and provided a productive and comfortable workplace for me to get my writing done. Thank you!

It's because of the professional skills of Dr. Mary Eileen Kungl that my cancer was diagnosed immediately. I'm not sure a "thank you" covers this. But—thank you. Grateful thanks also to OBGyn Dr. Christine Doyle, who went the extra mile, and to Dr. Barbara Buttin, my patient and thorough surgical oncologist. Grateful thanks to Christine McMinn, my oncological counselor, and the Living Well Cancer Center in Geneva, Illinois. My diagnosis and recovery is a tribute to their good care, and I am grateful to be in their hands going forward.

Many thanks to my adult children, Dustin Crosby and Jennifer Buono, and their spouses, Gillian Crosby and Nino Buono. They patiently put up with my tendency to insert dragonflies and prairie into a conversation about anything, and they send me photos of dragonflies they discover. So thankful for you all.

I couldn't chase dragonflies and write without the endless support of my best friend and husband, Jeff, who has hiked many a wetland and prairie with me and often sees a dragonfly or damselfly I've overlooked. His support of my writing and his concrete suggestions are part of every book I've written. JJ, you're my other set of eyes in the field, and you have my heart forever.

And lastly. Much love to my little dragonfly chasers, grandchildren Ellie, Jack, Tony, Anna, Emily, and Margaret, who help me see the natural world with fresh eyes each day. I hope the world holds nothing but awe and wonder for you. I feel confident leaving it in your hands.

◾

DESIGN
Marianne Jankowski

TYPESETTING
Marianne Jankowski
Steve Straus

PRODUCTION
Morris (Dino) Robinson

EDITORIAL
Anne Gendler
Mary Klein
Lori Meek Schuldt

TYPEFACES
Berthold Bakerville Book, Eccentric Std,
Eccentric Light and P22 FLLW Terracotta

PAPER, PRINTING & BINDING
70# White Opaque, 378 ppi
Integrated Books International, Dulles, Virginia

◾